26 NOV 1991

Indoor air quality:
biological contaminants

WHO Library Cataloguing in Publication Date

Indoor air quality : biological contaminants : report
    on a WHO meeting, Rautavaara, 29 August-2 September 1988

(WHO regional publications. European series ; No.31)

  1.Air quality 2.Air pollutants — adverse effects
  3.Allergens — adverse effects   4.Environmental health
  I.Series

ISBN 92 890 1122 X   (NLM Classification: WA 750)
ISSN 0378-2255

World Health Organization
Regional Office for Europe
Copenhagen

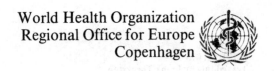

WHO Regional Publications
European Series No. 31

Indoor air quality: biological contaminants

Report on a WHO meeting

Rautavaara
29 August - 2 September 1988

ICP/CEH 073

Text editing by Frank Theakston

ISBN 92 890 1122 X
ISSN 0378-2255

PRINTED IN DENMARK

CONTENTS

Page

Introduction 1

Nature and definition of morbidity effects 4
    Terms and mechanisms 4

Hazard assessment 9
    Assessment of health outcome 9
    Assessment of hazard associated with environmental exposure 11
    Allergic mechanisms 13
    Irritants derived from bio-aerosols indoors 14
    Assessment of specific health hazards 15
    Inhalation of mycotoxins 17

Environmental measurement, sampling and analysis 18

Sources of microbiological contaminants 20

Environmental and biological factors in the indoor environment 24
    Ventilation 24
    Moisture 25
    Temperature 25
    Building microbiology 25
    Biologically derived particles from animals 26
    Pollen and fungi 27
    Bacteria and viruses brought indoors 27

Strategies for control of biological contaminants 28
    Building design and construction materials 29
    Ventilation and filtration systems 32
    Humidity control and air conditioning 32
    Other water systems 33
    Cleaning, maintenance and repair 33
    Behavioural factors 33
    Social, economic and regulatory considerations 35
    Guidelines, standards and regulations 37

Conclusions and recommendations 38

References 39

Annex 1. Assessments of levels of knowledge about
              indoor air quality 43

Annex 2. Membership of subgroups 51

Annex 3. Participants 52

Summaries in French, German and Russian 54

INTRODUCTION

Although it was recognized that not all the required or desired data are yet available, a Working Group on Indoor Air Quality: Biological Contaminants was convened from 29 August to 2 September 1988 in Rautavaara, Finland. The meeting was held in collaboration with the Department of Environmental Hygiene and Toxicology at the National Public Health Institute in Kuopio, and with the financial assistance of the Finnish Ministry of the Environment. It was attended by 19 temporary advisers from 11 countries, as well as by a representative of the Commission of the European Communities. Dr M.D. Lebowitz was elected Chairman and Dr J.A.J. Stolwijk Rapporteur. Dr M.J. Suess acted as Scientific Secretary.

The objective of the Working Group was to consider the state of knowledge about suspended viable particles, suspended allergens and other biologically derived suspended material. The Working Group focused on buildings and building elements and systems as sources of these biological air contaminants and on the resulting exposure of their occupants. The Working Group also considered the type, magnitude and distribution of the impact of these biological contaminants on the health and wellbeing of the occupants.

Much attention has been devoted to pathogenic organisms that multiply in buildings and building systems, such as *Legionella pneumophila* which causes Legionnaires' disease (legionellosis) and non-pneumonic Legionnaires' disease (Pontiac fever). Factors in buildings such as crowding and recirculated ventilation air can also promote the spread of airborne pathogens emitted by occupants suffering from tuberculosis, measles, varicella and other diseases.

Ventilation system components such as cooling towers, air chillers and humidifiers and dehumidifiers can support the growth of fungi, bacteria and other microorganisms. These can also grow on the structural parts of a building if the relative humidity inside the building reaches 70% or more, while dust mites can multiply in furnishings. Such microorganisms or products excreted by them or by arthropods and larger animals can be introduced into the indoor air, causing a variety of allergic and irritant reactions in the occupants.

1

Excessive concentrations of water vapour and accompanying condensation, water leaks, failures of equipment drains or lack of cleaning and maintenance can all contribute to the introduction of viable or biologically derived contaminants into building ventilation air. A sizeable proportion of the population has been, or is capable of being, sensitized over a lifetime to these forms of biological air contaminants. Other allergenic and toxic contaminants are animal dander, fragments of mites and other arthropods, and aerosols formed from animal faeces and urine.

The combined effect of all the biological air contaminants in indoor air is thought to account for a substantial proportion of absenteeism in schools and workplaces and of the days where activity is restricted. In the general population, five to ten days of restricted activity per head per year is a normal average. By reducing biological air contaminants indoors, acute infections and allergic episodes could be significantly reduced. It was pointed out that, in any building, the cost of losses in productivity due to absenteeism and restricted activity far exceeds the total cost of operating and maintaining the heating, cooling and ventilation systems. Because of the usual division of responsibility and authority in organizations occupying buildings, the relationship between these costs is not often considered.

Viable particles suspended in air, and other biologically derived particles in indoor air, form a distinct class of contaminant. When they are present in indoor air, even in small quantities, they can have a powerful effect on occupants. This effect can be through infection of the occupant by a suspended infectious agent, in which case the organism multiplies in the new host and can produce illness. There can also be allergic or irritant effects characterized by reactions ranging from uncomfortable to disabling. The Working Group met to consider the state of knowledge about the nature and magnitude of the adverse effects of biological air contaminants on human health and wellbeing, and to evaluate the role of buildings, building systems and contents in producing and disseminating such pollutants.

Some infectious diseases such as tuberculosis, Legionnaires' disease or measles have been demonstrated to spread through the airborne transmission of the infectious agent (1) and airborne transmission may play a contributory role in others. In many allergic conditions, building occupants can develop a hypersensitivity to secretions or to fragments or other products derived from animals and plants. Viruses do not multiply in buildings, but may spread from human and a few animal sources. Buildings can contribute to the airborne spread of viral disease, either through overcrowding or by the spread of airborne viruses through the ventilation system. Lack of an adequate supply of outdoor air will also increase the likelihood of the airborne transmission of infectious disease, through an increase in the concentration of suspended viruses and bacteria in droplet nuclei.

Exposure to airborne biological contaminants contributes to morbidity in the population. In the United States acute and chronic morbidity are continually surveyed, and for the year 1986 were reported as shown in Tables 1 and 2.

Table 1. Age-standardized measures of health status in the United States for 1986

Measure of health status	Number of days per person per year	Number of person-days per year (millions)
Days of restricted activity	15.10	3 624
Bed-days	6.50	1 560
Workdays lost (>18 years)	5.50	715
Schooldays lost (5–17 years)	5.00	400

Source: National Center for Health Statistics *(2)*.

Table 2. Prevalence of selected conditions in the United States in 1986

Condition	Prevalence per 1000 population
Heart disease	65.60
Hypertension	122.60
Cerebrovascular disease	11.90
Chronic sinusitis	145.50
Asthma	41.00
Migraine headache	36.00
Dermatitis	36.10
Allergic rhinitis	91.80

Source: National Center for Health Statistics *(2)*.

3

Table 2 reports on the prevalence of several conditions in the American population. Because any substantial role of the indoor environment in the causation of these conditions has not been considered until the last few years, there are virtually no public health surveillance data that allow estimates of the magnitude of this role.

To reduce exposures to harmful biological contaminants, we need to have information on the following:

— the nature and/or species of the suspended viable or biologically derived particles;
— the nature and mechanism of the morbidity effects associated with the contaminants, including the range and distribution of sensitivity in the population;
— the sources of the contaminants, multiplication sites, reservoirs and methods of distribution; and
— the characteristics of buildings and building elements and systems that affect the likelihood and degree of exposure to harmful biological contaminants.

NATURE AND DEFINITION OF MORBIDITY EFFECTS

There is a wide variety of biological agents and biologically derived materials in the indoor environment, and these are associated with a range of illnesses. In presenting these agents and their associated illnesses, it is important to rate their frequency and severity as well as the contribution to total incidence attributable to indoor exposure.

Each of the diseases or disease groupings presented in Table 3 was rated for severity and frequency on a scale presented in the footnotes. Table 3 presents a listing of the diseases and the agents, with a rating for each link between a disease and all the agent categories; this is also clarified in the footnotes to the table.

Terms and mechanisms

There are many diseases that have been associated with problems of indoor air quality. The following may sometimes be associated with or caused by biologically derived aerosols, as well as having causes unrelated to the building environment. The first four conditions below will be grouped together since the role of biologically derived aerosols is similar for each. They may be acute or chronic, and may deteriorate or improve in relation to biological aerosol exposure.

4

Rhinitis involves the nasal mucosa, and often results in itching or sneezing, and running or blocking of the nose. It may be due to infection (the common cold), allergy (e.g. hay fever) or nonallergic causes such as dryness or coldness of the air or air pollutants. It is commonly related to the indoor environment.

Sinusitis involves the sinuses. It causes pain or fullness in the face and may be associated with headache. It has causes similar to those of rhinitis.

Otitis is inflammation of the lining of the external (otitis externa) or internal (otitis media) parts of the ear. It may cause pain, and otitis media impairs hearing. The condition may be due to infection, allergy or nonallergic mechanisms.

Conjunctivitis involves the conjunctival mucosa. It causes itching, soreness, watering and discharge from the eyes. Its causes are similar to those associated with rhinitis.

Pneumonia is an infection of the gas-exchanging part of the lung resulting in consolidation of the lung. The condition is acute, and may be fatal. Most cases are unrelated to indoor air quality. Legionnaires' disease is a pneumonia caused by a specific bacterium (*Legionella pneumophila*). It accounts for less than 5% of community-acquired pneumonia. *Legionella* pneumonia may be building-related in about 30% of cases, mainly in hotels and hospitals.

Asthma is variable airways obstruction with bronchial irritability. It may be precipitated by biologically derived aerosols. Most of those affected have multiple trigger factors, such as exercise and cold air, irritants such as smoke and particulates, and allergens and drugs. Asthma symptoms related to particular buildings may deteriorate within minutes or hours of exposure, and may improve after leaving the building. Asthma may occur or reoccur 6–12 hours after exposure to an allergic stimulus in a building. This represents the late phase of dual-phase asthma, in which inflammatory factors may lead uniquely to a late-phase response or to a dual response commonly seen after exposure to industrial allergens or to other allergenic stimuli such as house dust *(5,6)*. Virus infections can precipitate acute attacks of asthma and may sometimes be the initiating cause.

Alveolitis is an inflammation of the gas-exchanging parts of the lung resulting in breathlessness. It may be acute, coming on about 4–12 hours after exposure. Chronic exposure may lead to permanent lung damage. It may be caused by allergic or nonspecific mechanisms. In the present context only allergic alveolitis will be considered.

Most building-related alveolitis is caused by contaminated humidifiers, containing many fungi and bacteria. In most cases the disease appears to be caused by soluble products rather than single whole organisms. Individual outbreaks have been related to any of the following organisms.

5

Table 3. Relationship between agents and disease with respect to severity, frequency and rate

Disease	Severity[a]	Frequency[b]	Agents[c]									
			Viruses	Bacteria (including endotoxins)	Fungi	Pollen	Animal dander and excretions	Mites[d] and excretions	Other arthropods	Amoebae	Actinomycetes and thermophilics	Plant constituents
Rhinitis Sinusitis Otitis Conjunctivitis	2	4	4 (infections)	0	3 (allergens)	3	4	1-4	2	0	1	1
Pneumonia	2-5	2	2	2	1	0	0	0	0	0	0	0
Asthma	2-5	3	3	0	2	4	4	1-4	2	0	1	0
Alveolitis	3-5	2	0	3	3	0	3	0	1	0	1-2	0
Humidifier fever	2-4	1-2	0	3[e]	3	0	0	0	0	0	1-2	0
Bronchopulmonary aspergillosis	2-5	1	0	0	4	0	0	0	0	0	0	0

Contact dermatitis	1–4	2–3	0	1	2	2	1	0	1
Atopic eczema	1–4	3–4	0	1	2	2	1–2	0	1
Contact urticaria	1–4	2–3	0	1	2	2	1	0	1
Mycotoxicosis	1–4	1	0	4	0	0	0	0	0

[a] Severity: 1 = trivial; 2 = interfering; 3 = restricting activities; 4 = incapacitating; 5 = serious, fatal.

[b] Frequency: 1 = rare; 2 = low frequency; 3 = medium frequency; 4 = common.

[c] Attributable rate: 0 = none; 1 = rare; 2 = low; 3 = medium; 4 = high.

[d] House dust mites as a cause of asthma and rhinitis relates largely to climate. They are found predominantly in humid environments; they are rare in desert and cold environments.

[e] Pontiac fever is a specific type of humidifier fever caused by various bacteria including Legionella pneumophila.

Source: Rom. W.N., ed. (3) and Turner-Warwick. M. (4).

Thermophilic bacteria	*Merulius lacrymans*
Micropolyspora faeni	*Trichosporon cutaneum*
Cytophagia allerginae	*Flavobacterium* spp.
Aspergillus fumigatus	*Bacillus subtilis*
Penicillium spp.	*Aureobasidium* spp.
Cephalosporium spp.	*Thermoactinomyces vulgaria*

Humidifier fever is an influenza-like illness developing 4–8 hours after exposure to aerosols from microbiologically contaminated humidifiers. Recovery occurs within 1–3 days despite continuing exposure. It classically occurs on the first day of re-exposure after a break of one or more days. It is often associated with headache and fatigue. Antibodies are found to *Acanthomoeba polyphaga* and *Naegleria gruberi*. The current evidence is that the antibodies are cross-reacting, and not the cause of the disease. Endotoxins are suspected to be the dominant cause of occupational cases.

Bronchopulmonary aspergillosis is a complicated specific form of asthma due to allergy to the fungus *Aspergillus fumigatus*. It can cause acute blockage of major airways and lung infiltrates. It is rare. The same fungus can cause asthma, rhinitis and alveolitis, may produce mycetomas in scarred lung tissue, or may be invasive in immunocompromised patients.

Contact dermatitis is an acute or chronic inflammation of the skin of variable severity due to allergic, toxic or irritant effects. Most causes relate to physical contact, but aerosols may be the cause (airborne contact dermatitis) *(7)*.

Atopic eczema is a chronic relapsing itching skin rash, variable in expression in genetically predisposed individuals. It commonly first occurs in infancy or early childhood. It is sometimes aggravated by biologically derived aerosols.

Contact urticaria is an acute or chronic, itching skin rash with variably sized wheals and swelling. It has allergic and nonallergic causes, sometimes caused by biologically derived aerosols *(8)*.

Mycotoxicosis is a rare toxic response to products from certain moulds, producing fatigue and irritability and inflammation of the heart. One building-related outbreak was associated with *Stachybotrys atra* [a] but there are many mycotoxin-producing species that need to be considered *(9–13)*.

The sick building syndrome consists of a number of symptoms that are common in the general population, but may in a temporal sense be related to a particular building. A substantial increase in the prevalence of the symptoms above background levels provides the link between the building and its occupants.

The main symptoms are *(14,15)*:

— eye, nose and throat irritation

---

[a] **Jarvis, B.** *Potential indoor air pollution problems associated with macrocyclic trichothecene by producing fungi.* Working paper available to the Health and Welfare Canada Working Group on Fungi and Indoor Air, 1986.

8

- sensation of dry mucous membranes, skin erythema
- mental fatigue, headache, nausea, dizziness
- high frequency of airway infection and cough
- hoarseness, wheezing, unspecified hypersensitivity.

The causes of sick building syndrome are many. It is epidemiologically related to sealed buildings, non-openable windows, tight-enclosure dwellings, increased temperature and dust levels, and passive cigarette smoking. There is also a likely role for biologically derived aerosols.

Allergy is an undesirable physiological event mediated via a variety of immunological mechanisms induced by specific allergens *(8,16)*.

Pseudo-allergic reactions are similar reactions without immunological specificity. They may be caused by the direct release of mediators, direct complement activation, psychoneurogenic effects or enzyme defects *(17)*.

There is epidemiological evidence that more workers have symptoms in buildings with humidifiers and chillers than in those without such equipment *(18)* and that very dry air (relative humidity less than 30%), which is common in heated premises in very cold climates, increases many sick building symptoms *(19)*. Humidifiers and chillers may become contaminated with microorganisms and are an important potential source of them. Experimental exposure of individual symptomatic workers to humidifier antigens can induce headache, rhinitis and lethargy, as well as asthma and alveolitis; similar exposures do not cause symptoms in previously unexposed workers.

HAZARD ASSESSMENT

Assessment of the effect of biologically derived aerosols has to be made for the medical management of individual patients, or for populations in the case of an outbreak in an exposed population. The connection between exposure and outcome is often not easy to make. Since the overall process usually begins with a presenting clinical outcome either in an individual or in a population, the first discussion will concern itself primarily with outcome assessment.

Assessment of health outcome

Individuals possibly exposed to bio-aerosols

The physician evaluating patients presenting with problems associated with indoor air quality is usually confronted with conditions affecting the upper and lower respiratory systems or the skin. More obscure symptoms affecting the central nervous system, including headache, drowsiness and

9

dizziness, may reflect psychosocial or psychological effects. Other systems, including the gastrointestinal, cardiovascular, endocrine and musculo-skeletal systems, are rarely affected.

The physician, epidemiologist or engineer, on being confronted with complaints from more than one occupant of a building, should visit the site to determine the pattern of illness. In office buildings attention should be paid to the application of hygiene standards. The next assessment should be of the ventilation system, with attention paid to air intakes and the type of air conditioner and humidifier. The rate of fresh air exchange or changes in the rate of exchange should be assessed. This may lead to measurement of several "indicator" contaminants, as well as total respiratory particulates. If allergy is suspected, sample collection for moulds at strategic sites including the ventilation system may be helpful. Carpets should be inspected for house dust mites, and the presence of pet animals or pests should be ascertained.

"Occupancy odour" due to the number of people in a room, resulting from a combination of many organic substances in low concentration, may be detected. Odourless conditions rarely exist in a building, with building odour consisting of a combination of factors, often including tobacco smoke, bathroom, waste and body odours, or cooking odours.

The medical history of the patient should include the onset of symptoms and their relationship to the building environment, including medical condition when away from the building, such as on weekends and holidays in the case of offices and on weekdays in the case of dwellings.

Physical examination of the patient should note obvious presence of conjunctivitis, the nature of upper airway change, any oedema or discharge, and dermatological changes such as rash, scaling, eczema or urticaria, no matter how trivial. This information would allow comparison of alternative work situations. The assessment, no matter what system is being evaluated, must indicate status both before and after exposure, including a monitored provocation assessment, the goal being to obtain all objective findings in relationship to the environment. This is easiest if the patient is asymptomatic prior to exposure, with evaluation at peak symptomatic periods. The work situation or environment may be "replicated" if specific substances such as cigarette smoke or cooking odour are suspected. These should be introduced singly or in combination in a controlled manner, such as in an environmental exposure facility.

Any objective documentation is most desirable, including photographs and pre- and post-exposure assessments, such as pulmonary function and upper airway resistance measurements by rhinomanometry. Other studies may include skin test evidence (patch or skin prick tests), culture of affected areas, and possibly biopsy. Precipitating antibodies to selected antigens or to natural mixtures of possible offenders should be measured for problems suggesting alveolitis. Chest X-ray on an individual basis may not only be helpful in diagnosis, but also for comparison. There are standard laboratory procedures for the various medical conditions, with additional tests for specific conditions.

Populations exposed to bio-aerosols

The investigation of populations requires techniques that differ substantially from those appropriate for the investigation of individual patients. Population studies depend for their validity on reliable epidemiological techniques, particularly the comparison of representative unbiased population samples exposed to different indoor air environments (14,18).

Questionnaires are the basis for investigation. They should be validated in the community to be studied in terms of comprehensibility, reproducibility and their power to identify the conditions under study. They should also be used to help define exposure, and to measure the major confounding factors. Questionnaire responses may be altered by the method of administration and the biases of the population being studied. It is easy to introduce biases if different interviewers are used. Some questionnaire responses are capable of validation under field conditions, such as tear film breakup time to validate eye dryness and lung function tests to validate asthma. Some responses are not easily capable of validation, such as lethargy and headache. Daily diaries of symptoms can be used as well to document longitudinal and time-specific responses in relation to exposure (20).

Exposure to antigenic material, either infectious agents or allergens, can be estimated by specific antibody determinations in populations. This is particularly appropriate for the study of exposure to *Legionella* spp., and for the study of sensitization to aero-allergens causing rhinitis, conjunctivitis and asthma (specific IgE estimations) and alveolitis or humidifier fever (specific IgG estimations) (1,3,4).

Peak expiratory flow rates can be obtained every 2–4 hours using a peak flow meter. This helps assess airway responses to exposure over time, and can elucidate the potential exposures of importance (for example in the evaluation of occupational asthma associated with chemicals). The response to quantified doses of allergens and irritants can be studied in individuals. In population studies, provocation of the eyes and nose can be studied easily, whereas bronchial provocation requires much more care. Tests should be carried out by experienced personnel, using controls and double-blind protocols. The resources required for such studies will be considerable.

Assessment of hazard associated with environmental exposure

Biologically derived aerosols produce their effects predominantly by infectious, allergic or irritant/toxic mechanisms. The health hazard

11

assessments differ with the different mechanisms and will be discussed separately.

Infectious mechanisms (21,22)

Pathogenicity is the ability of an infectious agent to cause disease in a susceptible host. This is not always an absolute characteristic of the microorganism, since those that are usually not pathogenic may cause disease under particular circumstances. In the case of viruses, however, pathogenicity is a more definite property of a given species.

Virulence is the degree of pathogenicity of an infectious agent, indicated by case fatality rates and/or its ability to invade and damage the host tissues. This may vary among different strains of the same species and subtype.

Although a dose–effect relationship exists (i.e. the more agents that are taken in, the more likely the host will become infected) the absolute or threshold infectious dose is not known for any of the bacteria that may be indoor contaminants. In the case of viruses, however, the presence of pathogenic species in the indoor air constitutes a health hazard in itself, because of the absence of a threshold.

Microorganisms need a favourable substrate to survive and reproduce in the environment and appropriate humidity and temperature. The concomitant presence of other moisture microorganisms may enhance or reduce their growth. Chemicals in the indoor environment, even if they are bactericidal or fungicidal, may favour growth through the positive selection of some strains. Viruses of importance as indoor air contaminants do not multiply outside the human body; they can survive in the indoor air for short periods of time but are not permanent contaminants.

The rate of emission of microbes into the indoor air depends on the nature of the source (e.g. high for cooling towers, low for house dust) and may be intermittent with high peaks. Emission of viruses depends on human behaviour such as sneezing and the emission of droplets during conversation (1,21). Transmission efficiency depends on the location of the source with respect to ventilation systems, air cleaning systems and air circulation. The presence of other suspended particles, solid or liquid, is very important since they are efficient carriers of microbes. Temperature and humidity influence transmission by altering particle size, thus affecting the settling time of airborne particulates (21,22).

Individual susceptibility to infections is of great importance in determining the actual risk of developing a clinical disease from exposure to viable agents in the indoor environment. Individual susceptibility may be increased under several conditions, the most important of which are:

— young age (particularly 0–3 years) and old age (over 60 years);
— existing disease, such as chronic obstructive lung disease;

12

- immunodepressive conditions such as occur under therapy with corticosteroids and other drugs, or during certain diseases such as cancer, AIDS and other chronic conditions;
- smoking and alcohol consumption and poor diet (i.e. low in necessary nutrients, vitamins and minerals);
- occupational or ambient exposure to airway irritants that may damage the pulmonary defence system.

Human susceptibility to infection decreases as a result of immunization (active or passive), including that due to prior infection. Active immunization is widely used against several infectious diseases such as tuberculosis, measles, rubella, mumps and pertussis. Immunity towards some infectious agents, either actively or passively acquired, can be measured by serum immunological tests determining the specific antibody titre, or by skin tests assessing the cell-mediated immune response. The presence of antibodies is indicative of exposure and infection, but it does not indicate when and where exposure occurred. This type of testing depends on availability of methods; accuracy is very important, especially if such testing is to result in probability distributions of immunity in populations *(22)*.

The detection of serum antibodies against *Legionella pneumophila* is of value in the diagnosis of Legionnaires' disease, but its significance as an indicator of immunity in healthy individuals is unknown.

Exposure assessment

Significant exposure to indoor infectious agents should be suspected when:

- there are several important sources
- the conditions favour microorganism survival
- there are complaints or epidemics of disease
- a microbiological laboratory reports many positive cultures or high rates of seropositivity.

Measurement of infectious agents can be made at the source, including humans, or in the air. These measurements are useful in identifying sources or the presence of specific agents. Quantitative assessment of exposure for individuals is difficult, and is of questionable value for many agents.

Allergic mechanisms

Two separate dose–response relationships exist. In a previously unsensitized population, the risk of sensitization is likely to depend on the potency of the allergen, the level of exposure and the length of exposure. Indoor air antigens vary in their potency, e.g. house dust mite antigens are often considered relatively potent and mould antigens less potent.

13

Current evidence suggests that a considerable percentage of the population is capable of developing IgE mediated sensitization (also referred to as atopy). The figure reaches around 60% of the population of Tucson, Arizona exposed to environmental allergens *(23)*. In occupational settings this figure may be higher, for example reaching around 75% in biological detergent workers exposed to antigens for *Bacillus subtilis (4)*. The whole population should therefore be considered at potential risk for allergy. Once sensitization has developed, only a proportion will express a related disease. Expression of disease depends on the dose of antigens, the level of antibodies in the individual and on non-specific amplification mechanisms (e.g. bronchial responsiveness for asthma, releasability of mediators, or skin reactivity).

Once sensitization has occurred, the dose–response relationship between exposure and disease differs from that of sensitization. Much lower doses are required to elicit disease in sensitized workers than to induce sensitization. Sensitized individuals who have become symptomatic will often respond similarly to other biological agents and to chemical agents such as formaldehyde and particles, due to heightened tissue reactivity *(3,4)*.

The control of allergic diseases in the population should concentrate on reducing sensitization, which can be measured by finding specific IgE antibodies. Using sensitization as the outcome measure, disease has been reduced in two occupational settings involving workers exposed to *B. subtilis* and laboratory workers exposed to rats. In both situations, exposure was assessed by measuring airborne antigens. Airborne antigen levels are likely to be the best measure of exposure; sufficient information exists to make this appropriate for some important indoor allergens, such as the Der P1 antigens from the house dust mite *Dermatophagoides pteronyssinus*. Mite dust allergens may be different in different areas.

Indoor air antigens differ widely in different climates. *D. pteronyssinus* is the most important allergen in western Europe, but becomes less prevalent at altitude and in northern Europe, where birch pollen becomes the most prevalent. Control of those allergens that cause rhinitis and asthma is also likely to control allergic alveolitis, a disease in which much larger doses of antigens are required to induce disease. The role of airborne biologically derived antigens is less clear in the development of skin diseases such as eczema, contact dermatitis and urticaria.

High levels of exposure to allergens for a short time is likely to result in more sensitization than a similar total dose over a longer period. Cumulative exposure is therefore a less relevant measure of exposure than recurrent peak levels *(4)*.

Irritants derived from bio-aerosols indoors

Bacteria growing in ventilation systems may liberate endotoxins, which are water soluble and may be distributed through the building via the ventilation system even in the absence of viable organisms. Most of our information comes from heavily contaminated sources. Endotoxin inhalation may cause an acute illness with fever, sweating, muscle aches, headache and sometimes rhinitis, asthma and breathlessness. Symptoms usually start 4–8 hours after exposure and resolve within 48 hours. Repeated exposures lead to tolerance and no further symptoms occur unless there is an interruption, usually of more than 24 hours after exposure. Endotoxin exposure is a possible cause of humidifier fever, and may be relevant to some of the symptoms of sick building syndrome or building-related illness. Control of bacterial growth will control endotoxin exposure of the occupants of a building. It is not known to what degree there are differences in individual susceptibility to endotoxin.

Fungi may produce mycotoxins that have potentially serious health effects. One outbreak of disease has been attributed to mycotoxins from *Stachybotryx atra*. A substantial growth of fungi is necessary to produce significant mycotoxin exposure.

House dust may produce its effects by non-specific irritant means, as well as by allergy to individual components such as house dust mites. Formaldehyde may play a similar role.

Little is known in real environments (i.e. outside animal exposure chambers) about possible interactions between biologically derived irritants and temperature, humidity, and other air contaminants such as inorganic particulates, nitrogen dioxide and ozone. Some individuals may be more sensitive to these other irritants. Many are definitely more sensitive to specific irritant effects or odours: up to 20% have eye hypersensitivity *(24)*. Those who are allergic often react more strongly to irritants as well, especially in the nose. Individuals with existing upper airway disease (up to 10% of the adult population) have an aggravated reaction to irritants and may show bronchospastic responses. Those with atopic eczema and dry skin also seem to be more sensitive to irritants.

Assessment of specific health hazards

Legionella pneumophila was first identified following an outbreak of pneumonia in Philadelphia in 1976 *(25)*. Since then, this organism has been implicated in a large number of outbreaks in Australia, Europe, and the United States and some in Asia, including some outbreaks that occurred before 1976 *(26,27)*.

Sporadic cases are reported with a frequency of 1–5% of all community-acquired pneumonias. The attack rate in common source outbreaks has been estimated at 2–10% (higher with increasing age) and the fatality rate is in the order of 10%. In Europe, around one third of the cases are associated with travel, and in several such events the source has been traced to a resort hotel.

15

Legionella spp. are ubiquitous bacteria in natural waters, and the commonly used methods of water treatment do not eliminate them. Thus they often enter buildings via the potable water system, probably in far too small numbers to cause infection. Amplification has been shown to take place in many locations in buildings such as cooling towers, domestic water heaters, water pipes, shower fittings and whirlpools, and also in some indoor equipment such as humidifiers. The optimum temperature for growth under laboratory conditions is 36°C, but growth will take place throughout the range 20–45°C; the presence of algae and sludge also promotes the growth of *Legionella* spp. Transmission is most likely to occur through inhalation of an aerosol created from, for instance, a shower tap or a cooling tower. In the latter case, the aerosol may reach people directly or via the ventilation system.

L. pneumophila is also associated with Pontiac fever, first identified in 1978 *(28)*. In comparison with the relatively low attack rate (2–10%) of Legionnaires' disease, its incubation period of 2–10 days, and its fatality rate of the order of 10%, Pontiac fever has a very high attack rate of 90% or higher and an incubation period of 12–36 hours. It produced no fatalities in the outbreak described.

Reported sources of *L. pneumophila* in different outbreaks have included cooling towers, humidification equipment and domestic hot water systems. Earlier WHO working groups have described Legionnaires' disease *(29)* and environmental aspects of its control *(30)*.

Enhanced spread of pathogenic airborne organisms through ventilation systems has been alleged or documented for a number of such pathogens *(1,31)*. Infected individuals are the source of the pathogens.

Early identification of disease is most important in order to prevent major outbreaks. This means that specific diagnostic procedures must be used, since clinically *Legionella* pneumonias do not differ significantly from those caused by other agents. A cluster (2 cases or more) of *Legionella* infections, especially if the organism has been isolated from a patient, should call for an investigation of any building with which the cases might be associated. It is then important that the site of amplification be located. The mere isolation of *Legionella* bacteria from a building does not, however, prove conclusively that the isolation site is the source of infection. Recently developed methods (monoclonal antibodies, isoenzymes) now allow subtyping of *L. pneumophila*, and it is important that such methods be used in epidemiological investigations.

Legionella bacteria are also frequently found in potable waters in buildings where no cases of Legionnaires' disease have been identified. The drinking of water contaminated with *Legionella* spp. has not been shown to cause disease, and inhalation of aerosols appears to be required. Current knowledge does not allow specification of the number of *Legionella* bacteria that can be allowed in potable waters without any risk of disease.

16

Isolation of *Legionella* bacteria from the air indicates a definite health hazard, regardless of how many are detected.

Mathematical models of epidemics describe well the spread of viruses through susceptible populations indoors, the accuracy of such models being a function of the size of the population. Models have been developed for measles, varicella and rubella that take into account their different viability out of the human environment, their passage through, for example, ventilation systems, and their incubation period, pathogenicity and virulence. Nosocomial infectious outbreaks due to these viruses, and due to *Pseudomonas* and *Staphylococcus* spp., have also been modelled. Unfortunately, these models have not been specified for different types of building, although attack rates and spread are more population oriented than they are building oriented *(32)*. One can assume that if susceptible populations exist, attack rate–time curves can be generated. These are compound epidemic curves, which continue to be compounded by the influx of new susceptibles. Each curve has parameters determined by exposure (dose), time, and the proportion of susceptibles. These compound distributions can represent the risk assessment exposure–response relationships, and can be defined for different populations.

Epidemic models of tuberculosis have been better defined in general and have been discussed for indoor environments *(32)*. Since tuberculosis occurs about ten times more frequently than Legionnaires' disease, the risk assessment models for different indoor settings in different populations should be defined so that risk management can proceed. Epidemic models have been discussed for certain outbreaks of Legionnaires' disease, but it does not occur frequently enough for general health hazard assessment.

Inhalation of mycotoxins

Another specific hazard concerns the inhalation of mycotoxins, i.e. toxic compounds produced by certain species of filamentous fungi. Mycotoxins are usually of concern with respect to ingestion of the substrate on which the fungus has grown, such as production of aflatoxin through the contamination of peanuts with *Aspergillus flavus*. However, the spores and mycelial fragments of toxigenic fungi contain high concentrations of toxins. Inhalation of the spores of the aflatoxin-producing *Aspergillus* spp., through harvesting corn or groundnuts or working in processing plants where these commodities are handled, has been shown to induce liver cancer. Inhalation exposure to certain mycotoxins has been shown to produce acute toxicity. It is very difficult to show that a mycotoxicosis is the underlying cause of patient distress; many mycotoxins induce subtle alterations in immune function. Exposure to abnormal concentrations of spores, mycelial fragments of toxigenic fungi or substrate particles should be considered hazardous *(9–13)*.

17

ENVIRONMENTAL MEASUREMENT, SAMPLING AND ANALYSIS

The selection and application of appropriate measurement, sampling and analytical strategies is central to the quantitative characterization of microbiological contaminants in non-industrial indoor environments. Such strategies are considered here from the viewpoint of patient diagnostic techniques and also from that of the building environment.

Physicians and health specialists are presented with two types of building-related symptom: those that are specific and can be studied directly, and those that are judged to be non-specific. In both instances the physician would follow the prescribed procedure of a complete medical history, with a physical examination and a series of diagnostic tests, in an effort to make a specific diagnosis. The medical history in such cases should include questions related to environmental factors, with particular emphasis on building-related factors. If the diagnosis indicates that specific microbial agents in the indoor environment may play an important role, a targeted environmental assessment to identify the source of the agents can be undertaken. For symptoms where indoor microbiological agents are suspected to play a role but no specific agents can be identified, a more general assessment of the building environment should be made.

The cornerstone of any environmental assessment strategy for investigating the presence either of specific microbiological agents (associated with specific symptoms) or of a more general mix of agents is a careful and thorough walk-through of the building. The walk-through would visually evaluate the presence of or the potential for microbiological contamination of the building or its systems. A particularly careful evaluation of the ventilation and humidification systems is required. Such an evaluation can identify potential sources of microbiological agents without the need for environmental sampling.

Where conditions warrant this, environmental samples are collected for laboratory analysis. If samples are required for patch provocation tests, they should be collected during the initial inspection, before any control measures are instituted.

Environmental sampling for microbiological agents in air and water and from surfaces should be used when the presence of such agents is suggested but no obvious sources can be identified, when confirmation of the presence of an agent is needed, or to rule out the presence of one or more agents. Selection of any environmental sampling method and subsequent laboratory analysis is determined by the medium to be sampled and the type of agents sought. There are no standard methods for the sampling and analysis of microbiological agents suspended in air, and thus care must be taken in designing a sampling protocol and in interpreting acquired data. The slow development of a standard sampling method has

18

resulted in delays in establishing a quantitative data base for the concentrations of microbial agents suspended in indoor air.

Collection of environmental samples can include air sampling, water sampling and collection of samples from surfaces or fabrics. Any such sampling programme has to address the potential spatial and temporal distribution of the biological agents, and be accompanied by detailed information about the site and the circumstances of collection. Particular care must be taken in collecting fungal spores because spore release is highly dependent on environmental factors such as temperature, humidity and light. In outdoor air seasonal variations may be of several orders of magnitude, and indoor samples should therefore be backed up with outdoor samples.

Air sampling can be conducted with either passive or active monitors. Passive monitors consist of growth medium or a sticky collection surface. The passive monitor can be used either as an area or personal monitor, and provides semi-quantitative information. Active air sampling consists of a pump that draws air over or through a collection surface, which can be growth medium or sticky material. Active samplers can be used as area or personal monitors. Air monitors that employ active systems can collect total biological samples (single filter) or be size selective (impactor). The collection of total aerosol mass and hence total biological agents permits the measurement of total biologicals; additional culture methods can then be used to obtain greater specification of the biological agents present. Impactors are used to provide a size distribution for the biologicals, and to permit the collection of specific biologicals directly on growth media. Filters can be eluted for direct tests of antigens in the air. Biological air samples can be collected on liquid impingers for subsequent laboratory analysis. Air and water samples and sediment samples should be accompanied by information on the physical properties (temperature, humidity, etc.) at the collection site. Surface scrapings should be accompanied by information about the type and quantity of surface area samples.

Selection of a growth medium for sample collection depends on the specific biological agent to be collected. For example, if an overall measure of high occupancy of an indoor space is desired, a general bacterial growth medium with a fungicide should be used. Analysis of collected samples consists of the macroscopic counting of colonies or microscopic counting of particles, plus possible further cultivation and characterization of the strains.

Evaluation of the results from environmental biological sampling has to be done with considerable care. The nature of the collection and analysis is highly dependent on differences in ecological conditions (climatological, seasonal and geographical), the growth medium used, the collection method, and laboratory procedures (incubation temperatures, incubation times, counting methods).

19

Reference samples in buildings, particularly in those without occupant complaints, should be collected to identify reasonable irreducible background levels. Owing to variations, the results from different investigators and laboratories cannot be directly compared. Efforts towards standardization are beginning to be made, and are much needed to allow such comparisons and to establish a data base on biological contaminants in different indoor environments.

The needs and means of exposure assessment are different for homes on the one hand, and offices and public buildings on the other. In homes, the people involved are few and permanent and the technical systems are relatively simple. In larger buildings the people involved include both temporary visitors or clients and more permanent employees, and the air handling and water systems, including their controls, are generally more complex. As a consequence, homes may be equipped according to special needs, for example those of allergic occupants. In most cases, sound construction techniques are more than sufficient. Office buildings are equipped according to more general guidelines. In homes the systems should be rugged and tamper-proof, and work reasonably well for long periods without maintenance. Usually they are adjusted and maintained only when they have broken down. The technical systems of office buildings are, at least in principle, regularly maintained by trained staff.

Climate and weather are crucial in the evaluation of exposure. In dry climates, such as a Scandinavian winter, humidifiers are a major potential source of allergens and bio-aerosols. In humid situations, such as maritime climates, mould and mite growth may be the most urgent indoor problem. Major environmental differences also exist between rural and industrial or urban areas that can affect the exposure situation inside buildings.

The factors affecting both source strength and removal processes vary widely with time. As a consequence, concentrations of viable aerosols and allergens in a given building are far from constant. This should be given adequate consideration when deriving exposure levels from measured data. Mite or animal allergens in dust samples can be detected by the ELISA (enzyme-linked immunosorbent assay) technique *(33)*. The guanine content of mite faeces can be measured by colorimetric reaction *(34)*.

SOURCES OF MICROBIOLOGICAL CONTAMINANTS

Microbiological materials are a major group of air contaminants in the nonindustrial indoor environment, and include moulds and fungi, viruses, bacteria, algae, pollens, spores and their derivatives. Microbiological contaminants indoors are generated by occupants, by building systems, by furnishings and their use, and by food. They also enter buildings in outdoor air. Measures to reduce or minimize human exposure to microbiological contaminants indoors are directed towards the sources or to the removal of the contaminants once airborne.

Given the wide range of microbiological agents, it is necessary to group them for the purposes of considering more generic source categories and control options. The following groupings will be considered:

- pollen, including the amorphous fraction and other plant constituents
- fungi and moulds, including hyphae, spores, conidia and mycotoxins
- mites, including fragments and excretions
- bacteria and viruses, including bacterial endotoxins, but only those that can survive outside the body and are transmitted through the air on dry surfaces or ventilation systems, not from person to person
- selected thermophilic bacilli such as *Legionella* spp.
- actinomycetes and thermophilics
- animal dander, including hair fragments
- excretions from insects and other arthropods, rodents, birds, cats, etc., that exist separate from the source
- amoebae.

Table 4 relates viable particles and biologically derived air contaminants to their respective sources in the indoor environment. Sources have been broken down into the following major categories.

Occupants comprise people, pet animals and pests.

Building elements and systems include all those sources that are an integral part of construction. Within building elements and systems 11 different sources have been identified. The humidification systems include both active and passive equipment, and centralized or local appliances. In the humidifier category an exception is found in high temperature steam injection systems, which so far have not been found to act as a source of indoor biological contamination. "Other water systems" comprise a variety of sources such as toilets, baths, sinks, drain traps, and aquariums which may give rise to biological aerosol emission due to their aeration systems. Organic building materials may degrade and thus become favourable substrates for microbiological growth. Of particular importance are casein-containing fillers used on floors and other surfaces, which are an optimal growth medium for actinomycetes.

Building use includes all those sources that may be related to internal equipment and use of the building. A very important source is represented by fabrics used for carpeting, furnishing, curtains, decorative articles, stuffed toys and cushions, and by any other naturally derived material such as straw or reed baskets. Besides these, other indoor surfaces independent of their constituent material may be contaminated by, and thus be a source of, microbiological contaminants.

21

Table 4. Biological contaminants in indoor air and their sources

Sources	Agents								
	Pollen	Fungi and moulds (including mycotoxins)	Mites	Bacteria and viruses (including endotoxins)	Bacilli	Actinomycetes and thermophilics	Animal dander	Animal excretions	Amoebae
Occupants									
People	—	—	—	2	—	—	1	—	—
Animals	—	—	—	2	—	1	3	1	—
Building elements and systems									
Moist construction materials	—	1	—	1	—	—	—	—	—
Air refrigeration units	—	2	—	1	—	—	—	—	—
Ventilation system	—	2	—	1	—	—	—	—	—
Evaporation coolers	—	1	—	1	1	—	—	—	—
Humidification system	—	3	—	1	2	3	—	—	2
Wet cooling towers	—	—	—	—	2	—	—	—	1
Water leaks	—	2	—	1	—	—	—	—	—
Other water systems	—	1	—	1	2	2	—	—	1
Moist basement	—	2	—	—	—	—	—	—	—
Deteriorating building materials	—	2	2	—	—	1	—	—	—
Other fibrous material	—	—	1	—	—	—	—	—	—

Building use

Fabrics	—	1	2	1	2	—	—
Other surfaces	1	2	2	1	—	3	—
House dust	2	1	—	1	3	—	3
House plants	1	1	1	—	—	—	—

Other

Outdoors	4	2	—	—	—	—	—
Food	—	1	1	—	1	—	—

Scale: 4 = predominant source; 3 = major source; 2 = important source; 1 = minor source; — = insignificant source.

23

The outdoor environment and food may be sources of indoor micro-biological contaminants. Food may act as a source because it is a multiplier for contaminants from other sources.

Although the symptoms associated with these agents are known, the direct association of sources of microorganisms and the symptom outcome reported is not a simple matter in any particular instance. However, the important role of microorganisms in one or more outcome is generally recognized. In a recent report, Morey *(35)* found that 18 of 21 buildings investigated after complaints by occupants contained microbiological reservoirs or amplifiers. The significant building factors reported by Morey in connection with such sources were, in order of their frequency of occurrence:

— inadequate maintenance of mechanical systems (13 buildings);
— mechanical system very difficult if not impossible to maintain (11 buildings);
— stagnant water in drain pans (10 buildings);
— porous man-made insulation (9 buildings);
— excessive relative humidity in the occupied space (6 buildings);
— floods in occupied space and air-conditioning system (6 buildings);
— outdoor air intake located near bio-aerosol source (6 buildings).

No information has been provided in the report quoted above as to whether the biological contamination was alleviated and whether, as a result, the symptoms have disappeared. It is important to note that several buildings had more than one source, and that in these buildings more than one outcome was reported. This multifactorial exposure associated with a multi-outcome effect is a common finding in buildings that develop an indoor air quality problem.

ENVIRONMENTAL AND BIOLOGICAL FACTORS IN THE INDOOR ENVIRONMENT

Ventilation

Indoor–outdoor air exchange will dilute indoor generated concentrations of viable air contaminants. It may also affect indoor humidity levels and thermal loads, and introduce viable outdoor contaminants. Mechanical ventilation systems can themselves become sources of viable aerosols. For example, air filters that are not routinely replaced can provide nutrients and strata for microbial growth.

Air exchange in winter reduces indoor moisture and can require the use of humidifying equipment. In summer, air exchange can increase indoor moisture, causing enhanced mould and mite growth, and dehumidification equipment may be required to avoid condensation on cold pipes and concrete slabs. Dehumidifiers may then become amplifiers of the biologicals.

24

The current ASHRAE standard 62-1989 *(36)* specifies minimum air exchange rates of 0.35 air changes per hour for dwellings or 7 litres per second per occupant, whichever is the greater, and 7 litres per second per person in other buildings. Depending on occupation densities, these requirements in general result in higher exchange rates in large buildings than in individual dwellings. Ventilation of dwellings at these rates in winter generally does not necessitate the use of humidificatin devices, whereas such devices may be required for large buildings.

In sealed buildings such as offices and schools, energy conservation measures may have reduced rates of indoor ventilation to the point where there is an excess of biological air contaminants from human sources. In such cases air velocity rates can be increased up to 0.2–0.3 m/s (the thermal comfort range). Ensuring adequate rates of air flow throughout the space will also assist in drying any materials unintentionally wetted as a result of leaks or condensation, thereby avoiding mould growth.

Moisture

Recirculated water sprays and stagnant waters in humidifiers and the drip pans of chillers and dehumidifiers can be significant sources of biological aerosols. While drip pans can be designed and installed so that they are properly drained by gravity, non-steam humidifiers must be continuously maintained or they will become an important source of biological aerosols. It is therefore best to control indoor humidity levels primarily through the regulation of ventilation rates, and the careful use of dehumidifiers, chillers and humidifiers where necessary.

Moisture level extremes in air can promote the survival of several species of microorganism and enhance the release of fungal spores by low humidity. As a result of all these factors, it is best that moisture levels be maintained in the 30–50% relative humidity range. At levels higher than 65% the incidence of upper respiratory illness might increase and adverse effects might occur in people suffering from asthma and allergies. Lower moisture levels (below 20%) may induce dryness or itching of the skin, and aggravate certain skin conditions.

Temperature

Human susceptibility to biological aerosols increases with warm temperatures. Perspiration associated with high ambient temperatures may increase eczema.

Building microbiology

Spores and viable bacteria are ubiquitous. In the indoor environment they form part of the house dust found on all surfaces, and in the air as aerosols.

In addition to these viable particles, endotoxins and mycotoxins are released into the indoor atmosphere by some of these microorganisms.

Once viable particles are present, the limiting factors of microbial growth are nutrients and moisture. Most construction materials, such as wood and particle board, and materials for sealing, filling and finishing contain adequate quantities and variety of organic compounds to support the start of microbial growth. Consequently, the factor that limits microbial growth in a building and its systems and surfaces is moisture.

Excess moisture or water may penetrate into the ceilings, walls, floors or furnishings through seepage, leakage or condensation of atmospheric water vapour. Microbial growth then starts immediately, leading to deterioration of building materials and the production of more viable aerosols in the indoor air. In air ducts, maintenance cleaning and treatment may be necessary. The remedy for excess moisture lies in adequate and appropriate design practices and effective maintenance. Removal of building material sources and the use of barriers are additional forms of control. A continuous problem, however, is the humid winter climate in certain areas so that water continues to collect and provide a growth medium for these microorganisms.

Water reservoirs associated with heating, ventilating and air conditioning are an especially favourable habitat for microbial growth. Reservoirs of this type can be found in the humidification or cooling systems in which the water is recirculated, allowing sediment formation. Aerated aquaria are another example of this type of reservoir. Any sediment in water is a habitat for a rich flora containing bacteria, fungi, actinomycetes, algae and amoebae. Depending on the temperature, the flora may be dominated by mainly thermophilic organisms. Any kind of mechanical disturbance of contaminated water may produce aerosols containing viable materials; evaporation alone does not appear to produce viable aerosols from contaminated water reservoirs. Regular sediment removal from water reservoirs is recommended to avoid microbial growth. Chlorination is recommended as a biocide in standing water, for example in the control of *Legionella* spp..

Humidification and water-based cooling systems can be maintained so as to minimize their contribution to bio-aerosols in the indoor environment. Chemical treatment can sometimes be one of the measures, but most chemicals produce their own hazards.

Indoor plants and the associated soil can support fungal growth, as can many foods. For control of such growth it is important to maintain such sources properly, and to remove food that has grown fungi.

Biologically derived particles from animals

Nonviable but biologically derived aerosols indoors consist of faecal matter from arthropods (especially house dust mites) and some insects, faecal

26

matter and urine from rodents, dander from people and animals (dogs or cats) and some potential parasites from animals. Depending on the region and its climate the composition of this contribution can vary according to the ecological niches present. These materials are found in house dust, surfaces, fabrics, moist building materials, pet enclosures and food. Dander may be brought in from outside on clothing, especially by people working with animals.

Pollen and fungi

Airborne pollen grains are mostly smooth, without spines and often round; they are covered by a thick and very resistant exine. The material inside, called sporopollenine, contains allergens that are excreted through apertures. Pollen can be produced by house plants, but most pollen found in the indoor environment originates outside. Pollen ranges in size from 5 to 100 micrometers. Fragments of pollen may be more allergenic than whole pollen, and may enter from outside more easily (37). Pollen from outside sources enters the indoor environment through passive or active ventilation, on clothing or on pets. Most pollen settles on the floor fairly rapidly, but may be resuspended through occupant activity. Pollen production outside is seasonal, and can also be very different in quantity and in dominant species, depending on climatic variation from one region to another and on the year-to-year variation in the local weather. Plant constituents from flowers can elicit contact dermatitis in exposed skin (7).

Bacteria and viruses brought indoors

Bacteria and viruses are usually brought indoors in human hosts (the major reservoir) and may be spread person-to-person; control of such spread is governed by traditional infectious disease practices (20,38). However, some bacteria (e.g. *Pseudomonas* spp. and *Mycobacterium tuberculosis*) and viruses (e.g. measles virus and human (alpha) herpesvirus 3) may be excreted and may stay viable on fabrics or surfaces ("foamites") or even get into ventilation systems (1). From these sources they may infect other humans. There are cleanliness and behavioural controls to avoid this secondary transmission. Ultraviolet radiation or chemical treatment may be used as a control measure, but may represent a hazard in itself.

Some microbiological agents enter the indoor environment as diseases of pets (e.g. toxoplasmosis in cats or rabbits and psittacosis in birds) and may be transmitted to humans. The main means of control is prevention of contact and the secondary means is removal of the source. Behavioural control methods are also important (see below). There are some animal viruses, such as cat leukaemia virus, that may be transmissible to humans but for which there is insufficient evidence (21).

STRATEGIES FOR CONTROL
OF BIOLOGICAL CONTAMINANTS

Strategies for the control of biological aerosols in indoor air are predominantly based on the avoidance of conditions that provide a substrate for the growth of viable particles, and secondarily on the containment and removal of such growth. Filtration of the air in a ventilation system can help reduce the number of biological aerosol particles, but is not a satisfactory primary strategy. Control of dew point in an indoor space is clearly the most important single way in which the growth of micro-organisms can be controlled, and many of the control strategies are aimed at that aspect. Aerosol formation involving water that may be contaminated is an important source of viable particles indoors, and cooling towers can be very large outdoor sources of bio-aerosols. Buildings with a high occupant density have an increased risk of airborne transmission of infectious diseases, and the presence of pets or pests can be responsible for important contamination of indoor air with nonviable allergenic aerosols. Table 5 sets out the control methods available for the various indoor sources. The importance of the sources and the effectiveness of the control methods are indicated, with a semi-quantitative rating system ranging from 1 to 4. In addition to the positive scores, negative scores have been used to indicate possible counteracting or secondary effects of the intervention.

Methods for source control have been placed in five groups. Because several of these controls can have the opposite effect for other contaminants, they should be used with caution.

- Proper design and construction of buildings is the most desirable means of avoiding biological pollution from buildings and building systems. This control strategy will be discussed below in more detail.
- The behaviour of the occupants of a building has been recognized to be a major determinant of the impact of indoor biological pollution. Consequently, appropriate behaviour is an effective way of achieving control of many sources, as discussed below in more detail.
- Source modification offers several opportunities for the control of biological pollution. Changes in temperature or relative humidity can be used to control some sources, but will often have the opposite effect on others.
- Maintenance, repair and cleaning are considered as a common group of control strategies. In this group, treatment with biocides and ultraviolet irradiation have also been included, but it should be recognized that the use of these methods should be strongly discouraged or limited because of the toxic risks for occupants. Because of these risks, these strategies have been given minus values in Table 5.
- The removal of pollutants from the air can be accomplished by

increasing the effective ventilation and/or by air cleaning. These methods, although in principle applicable to reducing emissions from all sources, are in practice of importance in only a few cases.

Building design and construction materials

Although operation and maintenance, occupant behaviour and regional climatic factors are important modifiers of the risk of developing biological contamination of indoor air, all these factors are influenced by the design and construction of a building. In turn, the major common factor on which the design and the choice of construction materials should concentrate is the effective control of moisture, which is the most important factor governing microbial growth in a building. Because of regional differences in climate, and because of regional variation in the availability of construction materials, in building practices and in craft skills, there will be different emphases in different regions in pursuing the control of moisture.

There are important differences in the design and construction of single-family detached residences and high-rise office blocks, and the critical features in the control of moisture will express themselves differently in such buildings. There are many instances where conflicts between energy conservation and effective moisture control can arise, or where the control of one source of moisture (e.g. sealing out of rainwater by a tight building envelope) will aggravate another source (e.g. moisture generated by cooking or by dense building occupancy rates). The following is presented in the recognition of all these interactions, and the comments and concerns should be considered in the light of local applicability.

The design of the structure should allow for the removal of any condensed moisture through adequate ventilation; ideally the structure should consist of non-deteriorating materials so as not to offer a substrate for microbial growth. The construction site should be dry, or should be so well drained as to become dry. The basement structure should be surrounded by a ground surface that slopes away from it, so as to minimize the possibility of rain run-off entering through cracks. If the basement is to be used for habitation, the enclosure should be finished so as to minimize the possibility of condensation problems on cold floor slabs, which could lead to mould growth. Floor drains should be trapped, and the water in the traps should be continuously flushed so that the trap does not become polluted. The walls, and especially the windows and window-framing, should be thermally insulated to prevent condensation on inside surfaces. The structure should be designed so that rainwater and melting snow cannot gain entry into the structure, even when the building and its materials age over the years. Siting of the structure and its air intakes should be such that it avoids

29

Table 5. Biological contaminants in indoor air:
sources and controls[a]

Sources	Controls			
	Behaviour			Prevention (design and construction)
	Education	Personal	Social	
Occupants				
People	2	3	3	—
Pets	1	4	2. - 1	—
Pests	1	2	3. - 1	1
Building elements and systems				
Moist construction materials	—	—	—	3
Air refrigeration units	2	1	1. - 1	1
Ventilation system	1	2	2	1
Evaporation coolers	2	1	1	1
Humidification system	2	2	—	1
Wet cooling towers	—	—	—	1
Water leaks	1	1	—	2
Other water systems	1	1	—	2
Moist basement	—	—	—	2
Deteriorating building materials	—	—	—	1
Other fibrous material	—	—	—	2
Building use				
Fabrics	2	2	1	—
Other surfaces	1	2	—	—
House dust	1	1	—	—
House plants	1	3	1	—
Others				
Outdoors	—	1	—	1
Food	1	1	—	—

	Controls (contd)				
	Source modification			Removal from air	
Temperature, relative humidity	Removal	Barriers	Repair, maintenance and cleaning	Ventilation	Air cleaning
—	—	—	1,[b] – 1	1	—
—	3	—	2, -1	—	—
1	—	1	1	—	—
1	1	2	—	1, -1	—
—	—	—	1, -1	—	1
—	1	—	1, -1	—	1
—	—	—	1, -1	—	1
—	1	—	1, -1	—	1
—	1	—	1, -1	—	1
—	—	—	4	—	—
—	1[c]	—	1	—	—
1	—	2	1	1	—
2	2	—	—	—	—
2	3	2	—	—	—
2	1	1	2	—	—
2	1	—	1, -2[d]	—	—
2	1	—	2, -2	—	1
—	2	—	1, -2	—	—
—	—	—	—	1, -1	1
1	—	3	1, -1	—	—

[a] *Scale:* 4 = highly effective; 3 = moderately effective; 2 = effective; 1 = slightly effective; -1 = detrimental; -2 = very detrimental; — = not applicable.

[b] Includes ultraviolet treatment.

[c] Aquarium.

[d] Application of termiticides.

31

air contamination by cooling tower drift, by exhaust air or by vehicle exhaust fumes.

Ventilation and filtration systems

The design and effectiveness of the ventilation and filtration system are of great importance. Simplicity and straightforwardness are desirable characteristics because of the necessity of ensuring that owners and operating personnel understand the system. Meters and gauges associated with the system should give understandable feedback to operating personnel. For systems with mechanical forced ventilation, each room should have provisions for a planned amount of ventilation air to be supplied, and the ventilation air should be distributed throughout the space so that all parts of the space are supplied equally. In residences without central mechanical ventilation it is necessary to provide efficient exhaust ventilation in kitchens and bathrooms, to remove moisture that is inevitably introduced into such spaces.

All components of the ventilation and filtering system should be easily accessible for cleaning and maintenance, and the ducts should also be accessible for cleaning. Ducts containing humid air should be as short as possible, should be in warm areas of a building, and should not run horizontally so that any condensation will collect and drain. It is very important that at the time of commissioning or transfer from builder to owner, or from one owner to the next, a complete set of design documentation and operating and maintenance instructions be transferred. Operating instructions should be appropriate for the operating personnel and should include instructions to seek professional assistance where necessary. Safe operation of a ventilation and filtration system should not have to rely on the routine use of biocides.

Humidity control and air conditioning

Humidity control and air conditioning are closely interrelated in most installations. Regional climates and the size and shape of the building dictate whether humidity control and air conditioning are required. In most instances humidification is required during the heating season when the outdoor water vapour pressure is very low. Preference should be given to humidification through the injection of steam that does not contain corrosion inhibitors and is delivered at temperatures above 65°C. Humidification systems that rely on the spraying or otherwise dispersing of water kept and discharged at temperatures below 65°C can easily become a major multiplier and disseminator of microbial contamination in a building. They require frequent cleaning and the use of biocides, which can have secondary effects. In some climates dehumidification devices may be required in basements for operation in the summer and autumn, when condensation is likely to occur on basement floor slabs.

Dehumidification and cooling of indoor air in cooling coils leads to condensation of water in such devices. This condensation must be able to drain and the overall system must be kept clean and free of microbial growth. This requires easy access to all the components on which water can condense and to the systems used for collection and drainage.

Other water systems

All spaces in which water is introduced and spilled should have waterproof floors, and should be well drained. This is of greatest importance in bathrooms, saunas and kitchens, but should also apply to basements. All plumbing should be installed so that it is protected from freezing, and all plumbing lines should be installed so that they can easily be inspected for leaks and repaired promptly when leaks occur.

Cleaning, maintenance and repair

During regular cleaning of a building the cleaning materials should be reviewed for their propensity to emit volatile organic compounds and other irritant or toxic materials.

Air ducts, air filters and other components of the ventilation system should be inspected at regular intervals, and a record should be kept of such inspections. All cooling towers, humidifiers and cooling coils should be inspected at regular intervals, and a cleaning schedule should be set up. The routine use of biocides should be avoided wherever possible. The functioning of all drains should be verified at every inspection. Buildings should be annually inspected for water leaks and water damage, whether from rainwater, from plumbing leaks or from water entering the basement through the soil. In case of water damage the damaged area should be thoroughly dried, and any damaged material should be removed and replaced.

Viable aerosol particles are present everywhere at all times, and any time a favourable substrate is encountered microbes will proliferate. Cleaning and disinfection are very important in the prevention and control of biohazards, but will provide only temporary relief if growth is not prevented by the design and construction and the choice of materials.

Behavioural factors

Numerous options are available to help control microbiological air contaminant levels in indoor air, such as source modification, prevention, repair, maintenance, cleaning and contaminant removal from air. It

33

may be, however, that modification of behaviour that influences indoor air contaminant levels and resultant exposure is often the simplest, least expensive and most effective means of improving indoor air quality and reducing adverse health effects. In the context of the microbiological agents considered here, the behavioural factors encompass a broad range of activities including education, personal behaviour, and social or cultural practices.

Education as discussed here is limited to the dissemination of information to the building occupants on action they could take to reduce and/or eliminate indoor microbiological agents. The ability to recognize situations that might contribute to concentrations of microbiological agents indoors and the action that could be taken by building occupants is, in large part, a function of the information available to the occupants. Information dealing with reducing or eliminating indoor microbiological agents has to be effectively transmitted to the public if it is to be used properly. For example, the public has to be made aware that the use of cool mist or ultrasonic room humidifiers could be a major source of microbiological agents, a source that could be reduced or eliminated if steam humidifiers were substituted. Health agencies, public health interest groups, the press, and professional and trade associations can play an important role in the dissemination of that information.

The availability of information, however, does not guarantee that the information will be used by the individual. Use of the information will depend on the motivation of the individual and available resources. Individuals could use the available information on source control or contaminant removal to reduce their exposures or remove themselves from a contaminated environment in which they experience adverse health effects.

Society can exert considerable pressure to control sources of contamination. For example, pressure to practise high standards of personal hygiene can reduce microbiological agents generated by humans. In addition, societal pressure can result in establishing adequate ventilation standards and ensure the institution of building maintenance practices that will minimize the potential for contamination by microbiological agents.

In the majority of buildings in which occupants have registered complaints, inadequate maintenance procedures have been found to contribute to the complaints (dirty make-up air intakes, missing or dirty filters, contaminated heating and cooling coils, contaminated drip pans, etc.). Employers of office personnel, building owners and those responsible for maintaining the building systems need to be educated on the importance of instituting maintenance practices to minimize air quality problems. Building designers need to be educated on the importance of designing air handling and humidification systems that permit easy access to those systems for routine maintenance. Groups responsible for establishing building

34

construction and maintenance guidelines or standards have to be sensitized to the importance of establishing and enforcing guidelines that will minimize microbiological contamination in buildings.

Social, economic and regulatory considerations

The incidence and prevalence of morbidity, as measured by days of restricted activity per person, is of the order of 10 – 20 days per year. It is difficult to estimate how much of this morbidity is due to biologically derived contaminants in indoor air. As indicated in Tables 1 and 2, the total impact in terms of absenteeism from school or employment is substantial. Milder forms of impact that do not result in absenteeism may still have substantial consequences in terms of lost productivity.

Different types of building may exhibit different levels of exposure to biological contaminants. These contributions are weighted by the amounts of time spent in such environments, as indicated in Table 6. This weighting should be considered when assigning priorities for control.

Table 6. Average exposures in different building types

Building type	Exposure (hours/day)
Residences	14
Offices	7
Public buildings	1
Schools	6
Outdoors and transportation	2

The likely frequency and severity of biological contaminants in the indoor air in different building types depends on geographic and climatic conditions as well as on construction style. The latter may vary from a total absence of heating, ventilation or air conditioning requirements to totally sealed climate-controlled buildings.

Although there are indications that biological contamination in buildings is very widespread, the exact determinants and distribution of biological contamination have not been systematically determined. Consequently, the financial burden associated with bio-aerosols and their control cannot be adequately estimated.

35

The relationship between the annual cost of heating, ventilation and air conditioning (HVAC) in a given building and the annual payroll cost of the employees served is such that the annual salary costs are 100 – 200 times the total HVAC costs, when both are measured in terms of unit of floor area. This suggests that even if the HVAC costs were to be doubled so as to reduce biological (and other) indoor air contaminants, only a very minor increase in productivity would be required to offset such increased costs.

A simplified example based on the experience of the General Services Administration (GSA) in the United States can be used to illustrate the quantitative relationships. In GSA operated buildings the average density of occupants is about one per 10 m^2. If we make the assumption that the average salary per occupant is of the order of US $25 000, then the costs of salary and building operation are as indicated in Table 7.

Table 7. Aggregate cost in US dollars of buildings operated by the US General Services Administration for the fiscal year 1987

	Total cost (US $ million)	Cost per m^2	Cost as percentage of average salary
Cleaning	173	7.52	0.277
Maintenance	121	5.26	0.194
Utilities/fuel	199	8.65	0.318
Other services	20	0.87	0.032
Administrative support	59	2.57	0.094
Total building management	572	24.87	0.915
Protective service	48	2.09	0.077
Administrative support	10	0.43	0.016
Total protection	58	2.52	0.093
Total direct cost	630	27.39	1.008

If the average absenteeism is 5 days per year (out of a total of about 200 working days annually), then this absenteeism represents a cost of US $62.50 per m². A reduction of absenteeism to 3 days per year would justify a doubling of the total cost of management of the building, even if productivity did not increase for the other 197 days in the year.

The costs of increased maintenance often fall on a building owner, or on firms that have contracted for the operation of a building, while the savings in absenteeism and the improvement in employee health would be enjoyed by the employer and employees. This separation of costs and benefits is a very effective barrier to instituting desirable improvements in operation and maintenance, and does not lead to optimization of the whole system.

In general, the return on investment in building-related modification to mitigate biological pollution will be high in terms of the financial benefits of reduced morbidity and increased productivity. The costs of these building-related modifications are likely to be small in comparison to current building costs.

Guidelines, standards and regulations

There is an increasing tendency in several countries for occupants who feel that they have been harmed by compromised air quality in buildings to sue architects, engineers and owners of buildings for compensation. To prevent such litigation, building codes or other relevant guidelines and standards should prescribe methods for the avoidance of biological contaminants in buildings and building systems. These can take the form of roposed policies and practices *(39)*. To protect the public, governmental guidelines for the reduction of biological contamination have been established *(40)*.

As another example, in its revised ventilation standard 62-1989 *(36)*, ASHRAE has included a number of requirements and guidelines to avoid the growth of biological contamination. One important requirement added in the new draft standard concerns easy access to ventilation systems.

In the case of filamentous fungi in indoor air, it has been agreed that specific numeric standards are, in practical terms, difficult to determine because the specific and relative hazards involved, as well as the sensitivity of individuals exposed, are variable. In Canada, residential air quality guidelines *(40)* call for exposure to microbials to be minimized as follows:

- the presence of certain fungal pathogens such as *Aspergillus fumigatus* and certain toxigenic fungi such as *Stachybotrys atra* should be considered unacceptable
- more than 50 colony forming units (CFU) per m³ of fungi from in-door sources should prompt investigation if there is only one species present

- up to 150 CFU per m³ of fungi from indoor sources should be considered acceptable if there is a mixture of species
- up to 500 CFU per m³ of fungi from indoor sources should be considered acceptable if the species present are primarily *Cladosporium* or other common phylloplane fungi; higher counts should be investigated to ensure there is no indoor source.

CONCLUSIONS AND RECOMMENDATIONS

Conclusions

1. A substantial portion of disease and absenteeism from work or school is associated with infections and allergic episodes caused by exposure to indoor air. Since this morbidity is often due to biological contaminants generated in buildings or to the crowding of occupants, it can be reduced significantly.

2. The increase in costs associated with improving the inadequate maintenance of ventilation systems results in greater comparative benefits in terms of better health for the occupants and reduced absenteeism.

3. Biological aerosols in buildings, including homes, are caused predominantly by persistent moisture and inadequate ventilation in spaces and building elements; proper design and construction are essential to prevent these conditions.

4. The levels of biological contaminants in indoor air vary enormously in time and space, so data bases on the distribution of the levels of contaminants in conjunction with occupants' response must be large enough to provide useful information for risk management.

5. Methods for collecting environmental samples of biological contaminants have generally not been standardized. Sampling methods for pollen, specific bacteria and viruses are close to standardization, but sampling methods for fungi, mycotoxins and other biological materials are not.

6. Laboratory procedures for the analysis of some fungi, mycotoxins, viruses, bacteria and other biologically derived materials of potential interest in indoor environments have not yet been standardized.

7. The use of biocides in the cleaning and maintenance of heating, ventilating and air conditioning systems or surfaces in buildings presents risks, both directly and through the promotion of resistant microbes.

38

Recommendations

1. Buildings and their heating, ventilating and air conditioning systems should not produce biological contaminants that are then introduced into the ventilation air. If the use of biocides is unavoidable, they should be prevented from entering space that can be occupied.

2. Standards and building codes should ensure the effective maintenance of ventilation systems by specifying adequate access and regular inspection and maintenance schedules.

3. In a building in which the occupants cannot effectively control the quality of ventilation air themselves, an individual who is responsible for this task should be made known to them.

4. To reduce allergic diseases in the community, total exposure to allergens should be minimized by controlling allergens and their sources in buildings.

5. For the risk assessment of allergic diseases, exposure–response curves should be established by measuring antigens in the air and specific IgE antibodies in the population.

6. Statistically designed population studies should be carried out using commonly accepted methods to obtain the concentration distributions of biological contaminants in specific geographic locations.

7. Sampling and analysis methods for aero-allergens and biological irritants should be standardized, and the effects of time, temperature and moisture should be determined.

8. The production of biological aerosols in buildings should be prevented by introducing appropriate prescriptions for design and construction practices into building codes.

9. The maintenance personnel of public and office buildings should be given adequate training in the routine inspection and maintenance of the buildings' systems.

10. Biological irritants and infectious agents cause nonspecific aggravation of respiratory and skin diseases, and they should be minimized by controlling their levels and sources in buildings.

REFERENCES

1. **Riley, R.L.** Indoor airborne infection. *Environment international,* **8**: 317–320 (1972).

2. *Current estimates from the national health interview survey, United States, 1987*. Washington, DC, US National Center for Health Statistics, 1988 (DHHS Publication (PHS) 88-1594).
3. **Rom, W.N., ed.** *Environmental and occupational medicine.* Boston, Little, Brown & Co., 1983.
4. **Turner-Warwick, M.** *Immunology of the lung.* London, Edward Arnold, 1978.
5. **Wasserman, S.** Basic mechanisms in asthma. *Annals of allergy*, **16**: 477–482 (1988).
6. **Booij-Noord, H. et al.** Immediate and late bronchial obstructive reactions to inhalation of housedust and protective effects of disodium cromoglycate and prednisolone. *Journal of allergy and clinical immunology*, **48**: 344–354 (1971).
7. **Benezra, C. et al., ed.** *Plant contact dermatitis.* St Louis, Mosby, 1985.
8. **Middleton, E. et al., ed.** *Allergy, principles and practice*, 3rd ed. St Louis, Mosby, 1988.
9. **Samson, R.A.** Occurrence of moulds in modern living and working environments. *European journal of epidemiology*, **1**: 54–61 (1985).
10. **Croft, W.A. et al.** Airborne outbreak of trichothecene toxicosis. *Atmospheric environment*, **20**: 549–552 (1986).
11. **Flannigan, B.** Mycotoxins in air. *International biodeterioration*, **23**: 73–78 (1987).
12. **Tobin, R.S. et al.** Significance of fungi in indoor air. *Canadian journal of public health*, **78**: 21–32 (1987).
13. **Sorenson, W.G. et al.** Trichothecene mycotoxins in aerosolized conidia of *Stachybotrys atra*. *Applied and environmental microbiology*, **53**: 1370–1375 (1987).
14. *Indoor air pollutants: exposure and health effects:* report on a WHO Meeting. Copenhagen, WHO Regional Office for Europe, 1983 (EURO Reports and Studies No. 78).
15. *Indoor air quality research:* report on a WHO Meeting. Copenhagen, WHO Regional Office for Europe, 1986 (EURO Reports and Studies No. 103).
16. **Ring, J. & Burg, G., ed.** *New trends in allergy II.* Berlin (West), Springer, 1986.
17. **Ring, J.** Pseudo-allergic reactions. *In*: Korenblat, P.E. & Wedner, H.J., ed. *Allergy, theory and practice*, 2nd ed. Orlando, Grune & Stratton, 1989.
18. **Finnegan, M.J. et al.** The sick building syndrome — prevalent studies. *British medical journal*, **289**: 1573–1575 (1984).
19. **Reinikainen, L.M. et al.** The effect of air humidification on different symptoms in an office building. An epidemiological study. *In: Proceedings of the Healthy Buildings Conference, Stockholm, Sweden.* Gavle, Swedish Council for Building Research, 1988, Vol. 3, pp. 207–215.

20. *Guidelines on studies in environmental epidemiology.* Geneva, World Health Organization, 1983 (Environmental Health Criteria No. 27).

21. **Benenson, A.S.** *Control of communicable disease in man,* 14th ed. Washington, DC, American Public Health Association, 1985.

22. **Fox, J.P. et al.** *Epidemiology.* New York, Macmillan, 1970.

23. **Barber, R.A. et al.** Longitudinal changes in allergen skin test reactivity in a community population sample. *Journal of allergy and clinical immunology,* **79**: 16–24 (1987).

24. **Weber, A.** Annoyance and irritation by passive smoking. *Preventive medicine,* **13**: 618–625 (1984).

25. **Fraser, D.W. et al.** Legionnaires' disease: description of an epidemic of pneumonia. *New England journal of medicine,* **297**: 1189–1197 (1977).

26. **Thacer, S.B. et al.** An outbreak in 1965 of severe respiratory illness caused by the Legionnaires' disease bacterium. *Journal of infectious diseases,* **138**: 512–519 (1978).

27. **Terranova, W. et al.** 1974 outbreak of Legionnaires' disease diagnosed in 1977. *Lancet,* **2**: 122–144 (1978).

28. **Glick, T.H. et al.** Pontiac fever: an epidemic of unknown etiology in a health department. I. Clinical and epidemiologic aspect. *American journal of epidemiology,* **107**: 149–160 (1978).

29. *Legionnaires' disease.* Copenhagen, WHO Regional Office for Europe, 1982 (EURO Reports and Studies No. 72).

30. *Environmental aspects of the control of legionellosis.* Copenhagen, WHO Regional Office for Europe, 1986 (Environmental Health Series No. 14).

31. **Zeterberg, J.M.** A review of respiratory virology and the spread of virulent and possibly antigenic viruses via air conditioning systems. *Annals of allergy,* **31**: 228–234; 291–299 (1973).

32. **Lebowitz, M.D. et al.** Health in the urban environment: the incidence and burden of minor illness in a healthy population. *American review of respiratory disease,* **106**: 824–841 (1972).

33. **Lind, P.** Purification and partial characterization of two major allergens from the house dust mite *Dermatophagoides pteronyssinus. Journal of allergy and clinical immunology,* **76**: 753–761 (1985).

34. **Bischoff, E. et al.** Zur Bekämpfung der Hausstaubmilben in Haushalten von Patienten mit Milbenasthma. *Allergologie,* **9**: 448–457 (1986).

35. **Morey, P.R.** Microorganisms in buildings and HVAC systems: a summary of 21 environmental studies (report on 21 buildings studied). *In : Engineering solutions to indoor air.* Proceedings ASHRAE IAQ88. Atlanta, ASHRAE, 1988.

36. *Ventilation for acceptable air quality.* Atlanta, ASHRAE, 1989 (Standard 62–1989).

37. **O'Rourke, M.K. & Lebowitz, M.D.** A comparison of regional atmospheric pollen with pollen collected at and near houses. *Grana,* **22**: 1–10 (1984).
38. *Communicable disease manual,* 15th ed. Atlanta, Centers for Disease Control, 1987.
39. **NAS/NRC Committee on Indoor Air Quality.** *Policies and procedures for control of indoor air quality.* Washington, DC, National Academy Press, 1987.
40. *Exposure guidelines for residential indoor air quality.* Ottawa, Health and Welfare Canada, 1987.

Annex 1

ASSESSMENTS OF LEVELS OF KNOWLEDGE
ABOUT INDOOR AIR QUALITY

During the meeting the Working Group reviewed the assessments made by previous Working Groups on Indoor Air Quality at meetings in Nördlingen in 1981, Stockholm in 1984 and Berlin (West) in 1987. These earlier assessments of the levels of knowledge had been presented in three tables: the levels of knowledge about population exposure in Table 1, the levels of knowledge about exposure–response relationships in Table 2, and a consensus on levels of public health concern based on current knowledge in Table 3. The present Working Group reviewed those parts of the prior assessments that dealt with contaminants of biological origin, without considering the other contaminants in the three tables. Thus the tables presented here have been revised only for biological contaminants.

Table 1. Current levels of knowledge
about indoor population exposure factors

Pollutant	People with low exposure	People with high exposure	Sources	Distribution	Instrumentation	Indoor and personal monitoring
Environmental tobacco smoke	most	some	+			±
Respirable suspended particles	most	few	±	0	±	0
Nitrogen dioxide	most	some	+	+	+	+
Carbon monoxide	most	few	+	+	+	+
Radon and daughters	some	variable[a]	+	+	+	+
Formaldehyde	most	few	+	+	+	+
Sulfur dioxide	most	few	+	+	+	+
Carbon dioxide	most	few	+	+	+	+
Ozone	most	few	+	+	+	+
Asbestos	most	few	+	±	+	±
Mineral fibres	most	few	±	0	±	0

Table 1 (contd)

Pollutant	People with low exposure	People with high exposure	Sources	Distribution	Instrumentation	Indoor and personal monitoring
Volatile organic compounds	most	few	±	+	+	+
Other organic compounds	most	few	0	0	±	0
Aero-allergens (pollen, fungi, biologically derived)	some	most	+	±	0	0
Infectious agents (bacteria/viruses)	most	few	+	±	±	0

[a] Distribution varies from area to area.

+ = adequate; ± = marginal; 0 = inadequate.

Table 2. Current levels of knowledge about exposure–response relationships

Pollutant	People with low exposure	People with high exposure	People exposed	Adverse effects at levels of concern	Exposure–response relationship	Means of control
Nitrogen dioxide	most	some	+	airway systemic	± 0	technical regulatory educational
Carbon monoxide	most	few	±	systemic	+	technical regulatory
Sulfur dioxide	most	few	±	airway	±	technical regulatory
Carbon dioxide	most	few	±	systemic	+	technical
Ozone	most	few	±	mucosal irritation airway odour	+ + ±	technical (indoors)
Radon and daughters	most	variable[a]	±	cancer	+	technical 'regulatory
Environmental tobacco smoke	most	some	±	odour irritation airway cancer systemic	+ + ± ± ±	technical regulatory educational social
Respirable suspended particles	most	some	+	mucosal irritation airway systemic	+ + +	

Table 2 (contd)

Pollutant	People with low exposure	People with high exposure	People exposed	Adverse effects at levels of concern	Exposure-response relationship	Means of control
Asbestos	most	few	0	cancer respiratory disease	+ ±	technical regulatory
Mineral fibres	most	few	0	irritation airway cancer	± ± 0	technical regulatory educational
Volatile organic compounds	most	some	+	odour sensory irritation mucosal irritation systemic airway cancer	0 0 0 ±[b] 0 ±[b]	technical regulatory educational social
Formaldehyde	most	few	±	odour mucosal irritation systemic airway cancer	+ ± ± ±	technical regulatory
Other organic compounds	most	few	0	odour mucosal irritation systemic airway cancer	0 0 ±[b] 0 ±[b]	technical regulatory educational social

Aero-allergens	some	most	±	mucosal irritation airway	± ±	technical medical[c] educational
Infectious agents	most	few	0	respiratory organs in other organs systemic	± 0 0	technical regulatory medical[c] educational

[a] Varies with region

[b] For some pollutants, knowledge is inadequate

[c] The medical measures are preventive

+ = adequate; ± = marginal; 0 = inadequate

Table 3. Consensus of concern about selected indoor air pollutants at 1987 levels of knowledge

Pollutant[a]	Typical range of concentrations reported[b] (mg/m^3)	Concentration of limited or no concern[b] (mg/m^3)	Concentration of concern[b] (mg/m^3)	Remarks
Respirable particles (including environmental tobacco smoke)	0.05–0.7	<0.1	>0.15	Japanese standard 0.15 mg/m^3
Nitrogen dioxide	0.05–1	<0.15	>0.3	AQG[c] value <0.4 mg/m^3
Carbon monoxide	1–100	2% COHb <11	3% COHb >30	99.9% of the concentration with continuous exposure[d]
Radon and daughters	10–3 000 Bq/m^3	carcinogen	carcinogen	Swedish standard for new houses 70 Bq/m^3
Sulfur dioxide	0.02–1	<0.35	>0.5	10 min SO_2 alone
Carbon dioxide	600–9 000	<1 800	>12 000	1 800 mg/m^3 is a widely used indicator
Ozone	0.02–0.4	0.1	0.12	
Asbestos	100–10^5 fibres/m^3	carcinogen	carcinogen	Optical microscopy >5μm
Mineral fibres	100–10^4 fibres/m^3	—[e]	—[e]	Skin irritation

49

Organic compounds		<0.06	>0.12	Valid for both long- and short-term exposure
Formaldehyde	0.02-2	carcinogen	carcinogen	
Benzene	0.01-0.04	carcinogen	3	AQG[c]
Dichloromethane	0.005-1	—[e]	1	AQG[c]
Trichloroethylene	0.0001-0.02	—[e]	5	AQG[c]
Tetrachloroethylene	0.002-0.05	—[e]		AQG[c]
1,4-dichlorobenzene	0.005-0.1	—[e]	450	TLV[f]
Toluene	0.015-0.15	—[e]	375	TLV[f]
m,p-xylene	0.01-0.05	—[e]	435	TLV[f]
n-nonane	0.001-0.03	—[e]	1 050	TLV[f]
n-decane	0.002-0.04	—[e]	—[e]	
Limonene	0.01-0.1	—[e]	560	TLV[f]

[a] All gases were considered on their own without other contaminants.

[b] Short-term exposure averages.

[c] AQG = _Air quality guidelines for Europe._ Copenhagen, WHO Regional Office for Europe, 1987 (WHO Regional Publications, European Series, No. 23).

[d] According to Environmental Health Criteria No. 13. Geneva, World Health Organization, 1979.

[e] No meaningful figure can be given because of insufficient knowledge.

[f] TLV (threshold limit values) established by the American Conference of Governmental Industrial Hygienists (1987/1988). These values are for occupational exposures and might be considered the extreme upper limit for non-industrial populations for very short-term exposures.

MEMBERSHIP OF SUBGROUPS

Subgroup 1 on health effects

V. Bencko
J.S.M. Boleij
S. Burge (*Rapporteur*)
J.H. Day (*Leader*)
I. Farkas

D.J. Moschandreas
B. Seifert
J.A.J. Stolwijk
D. Vieluf
D.S. Walkinshaw

Subgroup 2 on sources and controls

M.K. Hjelmroos
M.J. Jantunen
I.E. Kallings
P. Kalliokoski
H. Knöppel

B.P. Leaderer *(Rapporteur)*
M.D. Lebowitz *(Leader)*
M. Maroni
A. Nevalainen
H. Werner

Subgroup 3 on sampling and analysis methods

V. Bencko
J.H. Day
M.K. Hjelmroos

B.P. Leaderer (*Rapporteur*)
A. Nevalainen (*Leader*)
D. Vieluf

Subgroup 4 on exposure and health hazards assessment techniques

J.S.M. Boleij
S. Burge (*Rapporteur*)
M.J. Jantunen
I.E. Kallings (*Leader*)

P. Kalliokoski
M.D. Lebowitz
M. Maroni

Subgroup 5 on regulatory and economic considerations

I. Farkas
H. Knöppel
D.J. Moschandreas
B. Seifert (*Leader*)

J.A.J. Stolwijk
D.S. Walkinshaw (*Rapporteur*)
H. Werner

PARTICIPANTS

Temporary advisers

Dr Vladimir Bencko, Head, Department of Hygiene, Institute of Tropical Health, Postgraduate School of Medicine and Pharmacy, Prague, Czechoslovakia

Dr Jan S.M. Boleij, Professor, Department of Air Pollution, Agricultural University, Wageningen, Netherlands

Dr Sherwood Burge, Consultant Physician, East Birmingham Hospital, and Senior Lecturer, Institute of Occupational Health, Solihull, Birmingham, United Kingdom

Dr James H. Day, Professor of Medicine and Head, Division of Allergy and Immunology, Department of Medicine, Queen's University, Kingston, Ontario, Canada

Dr Ildiko Farkas, Head, Department of Experimental Hygiene, National Institute of Hygiene, Budapest, Hungary

Dr Mervi K. Hjelmroos, First Curator, Palynological Laboratory, Swedish Museum of Natural History, Stockholm, Sweden

Dr Matti J. Jantunen, Head, Environmental Hygiene Laboratory, Department of Environmental Hygiene and Toxicology, National Public Health Institute, Kuopio, Finland

Dr Ingegerd E. Kallings, Chief, Bacteriological Diagnosis Section, Department of Bacteriology, National Bacteriological Laboratory, Stockholm, Sweden

Dr Pentti Kalliokoski, [a] Professor of Environmental and Occupational Hygiene, University of Kuopio, Finland

Dr Brian P. Leaderer, Associate Fellow, John B. Pierce Foundation Laboratory, and Associate Professor, Department of Epidemiology and Public Health, Yale University School of Medicine, New Haven, CT, USA

Dr Michael D. Lebowitz, Professor of Internal Medicine and Associate Director (Environmental Programs), Division of Respiratory Sciences, College of Medicine, Health Sciences Center, University of Arizona, Tucson, AZ, USA (*Chairman*)

---

[a] Part-time participation

Dr Marco Maroni, Professor, Institute of Occupational Health, University of Milan, Italy

Dr Demetrios J. Moschandreas, Senior Science Advisor and Professor, Chemistry and Chemical Engineering Research Department, ITT Research Institute, Chicago, IL, USA

Dr Aino Nevalainen, Senior Research Scientist, Environmental Hygiene Laboratory, Department of Environmental Hygiene and Toxicology, National Public Health Institute, Kuopio, Finland

Dr Bernd Seifert, Professor and Deputy Director, Air Hygiene Department, Institute for Water, Soil and Air Hygiene, Federal Health Office, Berlin (West)

Dr Jan A.J. Stolwijk, Professor and Chairman, Department of Epidemiology and Public Health, Yale University School of Medicine, New Haven, CT, USA (*Rapporteur*)

Dr Dieter Vieluf, Assistant Physician, Department of Dermatology, Ludwig-Maximilian University, Munich, Federal Republic of Germany

Dr Douglas S. Walkinshaw, Coordinator, Indoor Air Quality, Environmental Health Centre, Health and Welfare Canada, Ottawa, Ontario, Canada

Dr Horst Werner, Head, Air Hygiene Inspection, State Hygiene Inspectorate, Ministry of Public Health, Berlin, German Democratic Republic

Commission of the European Communities (CEC)

Dr Helmut Knöppel, Head, Environmental Chemicals Sector, Chemistry Division, CEC Joint Research Centre, Ispra Establishment, Italy

WHO Regional Office for Europe

Dr Michael J. Suess, Regional Officer for Environmental Health Hazards, Copenhagen (*Scientific Secretary*)

RÉSUMÉ

Un groupe de travail composé de 19 conseillers temporaires venus de 11 pays ainsi que de représentants de la Commission des Communautés européennes et du Bureau régional de l'OMS pour l'Europe s'est réuni à Rautavaara (Finlande) pour faire le point des connaissances sur les particules, allergènes et autres matières en suspension d'origine biologique. Le groupe s'est concentré sur les bâtiments et leurs systèmes comme sources de ces contaminants biologiques de l'air, et sur l'exposition qui en résulte pour les occupants. Il a aussi examiné le type, l'ampleur et la répartition de l'impact de ces contaminants sur la santé et le bien-être des occupants.

Les participants se sont beaucoup intéressés aux organismes pathogènes qui se multiplient dans les bâtiments et leurs systèmes, par exemple *legionella pneumophila*, cause de la maladie des légionnaires (légionnellose) et la maladie des légionnaires non pulmonaire (fièvre de Pontiac). Des facteurs tels que le surpeuplement des immeubles et la remise en circulation de l'air de ventilation peuvent également favoriser la propagation des pathogènes transportés par l'air émis par les occupants souffrant de tuberculose, de rougeole, de varicelle et d'autres maladies.

Les composants des systèmes de ventilation, tels que les tours de refroidissement, les refroidisseurs d'air, les humidificateurs et déshumidificateurs peuvent favoriser la croissance de champignons, de bactéries et d'autres microorganismes; ceux-ci peuvent se développer également sur les éléments structurels d'un bâtiment si l'humidité relative à l'intérieur de ce dernier atteint 70% ou plus, alors que les acariens de la poussière peuvent se multiplier dans les meubles. De tels microorganismes, ou les produits excrétés par eux ou par des arthropodes ou de plus gros animaux, peuvent s'introduire dans l'air intérieur et provoquer chez les occupants diverses réactions allergiques ou irritantes.

Des concentrations excessives de vapeur d'eau et la condensation qui en résulte, les fuites d'eau, les défauts des canalisations ou l'absence de nettoyage et d'entretien sont autant d'éléments qui peuvent contribuer à l'introdduction de polluants biologiques ou d'origine biologique dans l'air de ventilation des bâtiments. Une proportion appréciable de la population est ou peut être sensibilisée, pendant la durée de son existence, à ces formes de

54

polluants biologiques de l'air. D'autres contaminants allergènes et toxiques sont les écailles d'animaux, des fragments de mites et d'autres arthropodes, et des aérosols formés de déjections et d'urine animales.

On considère que l'effet combiné de tous les contaminants biologiques de l'air à l'intérieur des locaux est à l'origine d'une forte proportion de l'absentéisme dans les écoles et sur les lieux de travail, ainsi que des jours où l'activité est réduite. Dans l'ensemble de la population, cinq à dix jours d'activité réduite par an constituent une moyenne normale. En réduisant les contaminants biologiques de l'air à l'intérieur des locaux, on pourrait réduire sensiblement les infections aiguës et les épisodes allergiques. Il a été souligné que, dans n'importe quel bâtiment, le coût des pertes de productivité dues à l'absentéisme et à une activité réduite était largement supérieur au coût total de fonctionnement et d'entretien des systèmes de chauffage, de refroidissement et de ventilation. Etant donné la répartition habituelle des responsabilités et de l'autorité au sein des organisations qui occupent ces bâtiments, les relations entre ces coûts ne sont que rarement prises en considération.

Conclusions

1. Une part importante des maladies et des absences au travail et à l'école est due à des infections et des épisodes allergiques causés par l'exposition à l'air intérieur. Cette morbidité étant souvent imputable à des contaminants biologiques produits dans les bâtiments, ou au nombre élevé des occupants, elle peut être sensiblement réduite.

2. L'augmentation des coûts liés à l'amélioration d'un entretien insuffisant des systèmes de ventilation a pour effet un accroissement des avantages comparatifs sous forme d'amélioration de la santé des occupants et de diminution de l'absentéisme.

3. Les aérosols biologiques dans les bâtiments, y compris les lieux d'habitation, ont pour causes principales une humidité persistante et une ventilation insuffisante des espaces et des éléments de construction; une bonne conception et une construction de qualité sont indispensables pour éviter une telle situation.

4. Les concentrations de contaminants biologiques dans l'air intérieur varient énormément dans l'espace et dans le temps, de sorte que les bases de données sur la distribution de ces concentrations et de la réaction des occupants doivent être suffisamment importantes pour fournir des informations utiles en vue d'une gestion des risques.

5. Les méthodes de collecte d'échantillons de contaminants biologiques dans l'environnement n'ont généralement pas été normalisées. Des méthodes d'échantillonnage sont près d'être normalisées pour le pollen,

certaines bactéries et certains virus, mais pas pour les champignons, les mycotoxines et autres matières biologiques.

6. Les procédures de laboratoire pour l'analyse de certains champignons, mycotoxines, virus, bactéries et autres substances d'origine biologique susceptibles de présenter de l'intérêt dans les espaces clos n'ont pas encore été normalisées.

7. L'utilisation de biocides pour le nettoyage et l'entretien des systèmes de chauffage, de ventilation et climatisation ou les surfaces des bâtiments présente des risques, à la fois directement et indirectement, en favorisant le développement de microbes résistants.

Recommandations

1. Les bâtiments et leurs systèmes de chauffage, de ventilation et de refroidissement ne devraient pas produire de contaminants biologiques s'introduisant dans l'air de ventilation. Si l'emploi de biocides est inévitable, il faudrait les empêcher de pénétrer dans des locaux qui peuvent être occupés.

2. Les normes et les codes de construction devraient assurer le bon entretien des systèmes de ventilation en précisant un accès adéquat ainsi que des calendriers d'inspection et d'entretien réguliers.

3. Les occupants d'un bâtiment, qui ne peuvent régler eux-mêmes efficacement la qualité de l'air de ventilation, devraient pouvoir s'adresser à une personne responsable de cette tâche.

4. Pour diminuer le nombre de maladies allergiques dans la collectivité, il faudrait limiter au minimum l'exposition totale aux allergènes en luttant contre ces derniers et contre leurs sources dans les bâtiments.

5. Pour évaluer les risques de maladies allergiques, il faudrait établir des courbes d'exposition-réponse en mesurant les antigènes présents dans l'air et les anticorps IgE spécifiques de la population.

6. Il faudrait effectuer des études statistiques dans la population en appliquant les méthodes communément admises pour obtenir la distribution des concentrations de contaminants biologiques en des endroits déterminés.

7. Il faudrait uniformiser les méthodes d'échantillonnage et d'analyse des allergènes et des irritants biologiques présents dans l'air, et déterminer les effets de la durée, de la température et de l'humidité.

8. Il faudrait empêcher la production d'aérosols biologiques dans les bâtiments en introduisant dans les codes de construction des prescriptions appropriées pour les pratiques de conception et de construction.

9. Il faudrait donner au personnel d'entretien des bâtiments publics et des immeubles de bureaux une formation adéquate dans les domaines de l'inspection et de l'entretien de routine des systèmes des bâtiments.

10. Les irritants biologiques et agents infectieux causent une aggravation non spécifique des affections respiratoires et dermiques, et il faudrait les limiter au minimum en luttant contre leurs niveaux et leurs sources dans les bâtiments.

ZUSAMMENFASSUNG

In Rautavaara, Finnland, tagte eine Arbeitsgruppe, der Berater auf Zeit aus elf Ländern und Vertreter der EG-Kommission sowie des europäischen WHO-Büros angehörten. Ziel der Arbeitsgruppentagung war die Feststellung des Wissensstandes über biologische, allergisierende und andere biologisch entstandene Schwebstoffe. Die Arbeitsgruppe konzentrierte sich auf Gebäude und Gebäudeeinrichtungen, in den bzw. durch die biologische Luftschadstoffe verteilt werden können, denen dann die Bewohner ausgesetzt sind. Die Arbeitsgruppe befasste sich auch mit der Auswirkung, die diese biologischen Schadstoffe auf Gesundheit und Wohlbefinden der Bewohner haben können.

Besonders beachtet wurden pathogene Organismen, die sich in Gebäuden und Gebäudeeinrichtungen vermehren. Dazu zählt die Legionella pneumophila, die die Legionärskrankheit (Legionellose) und die leichtere Form, das Pontiac-Fieber, hervorruft. Andere gebäudebezogene Faktoren wie überfüllte Räume und Luftumwälzung können ebenfalls zur Verbreitung von Krankheitskeimen in der Luft beitragen, die von Menschen ausgehen, die an Tuberkulose, Masern, Windpocken usw. erkrankt sind.

Verschiedene Teile von Klimaanlagen wie Kühltürme, Luftkühler, Befeuchter und Lufttrockner können das Wachstum von Pilzen, Bakterien und anderen Mikroorganismen fördern. Diese Erreger können sich auch auf eigentlichen Gebäudeteilen vermehren, wenn die relative Luftfeuchtigkeit in den Gebäuden über 70 Prozent beträgt; ausserdem steigt dann die Vermehrungsrate der Staubmilben in den Möbeln. Derartige Mikroorganismen bzw. ihre Exkrete oder die von Anthropoden bzw. grösserer Tiere können in die Innenraumluft gelangen und eine Reihe von allergischen Reaktionen oder Reizzuständen beim Menschen hervorrufen.

Hohe Wasserdampfkonzentrationen mit Kondenswasserbildung, Undichtigkeiten, fehlerhafte Wasserablassvorrichtungen und unsachgemässe Reinigung bzw. Instandhaltung können bewirken, dass lebensfähige biologische Schadstoffe in die raumlufttechnischen Anlagen geraten. Ein beträchtlicher Teil der Bevölkerung wird im Laufe des Lebens gegenüber solchen biologischen Luftschadstoffen sensibilisiert. Ebenfalls eine Rolle spielen Allergene und toxische Luftschadstoffe wie Tierhautschuppen,

58

Fragmente von Milben und anderen Arthropoden sowie aus Tierfäzes und -urin entstehende Aerosole.

Man vermutet, dass die biologischen Luftschadstoffe in der Innenraumluft in ihrer Gesamtheit für einen grossen Teil der Fehltage in der Schule und am Arbeitsplatz sowie für Aktivitätseinschränkungen verantwortlich sind. In der normalen Bevölkerung leidet jede Person an fünf bis zehn Tagen unter Aktivitätseinschränkungen. Durch Begrenzung der biologischen Schadstoffe in der Innenraumluft könnten akute Infektionen und allergische Anfälle signifikant reduziert werden. Es ist aufgezeigt worden, dass in bezug auf jedes Gebäude die durch Fehltage bzw. Aktivitätseinschränkungen erlittenen Produktionsverluste höher liegen als die Gesamtkosten für den Betrieb und die Instandhaltung der Heiz-, Kühl- und Belüftungsanlagen. Aufgrund der normalerweise anzutreffenden Kompetenzenteilung seitens des Gebäudebenutzers wird der Bezug zwischen diesen Kostenarten oft nicht gesehen.

Schlussfolgerungen

1. Krankheiten sowie Fehltage in der Schule und am Arbeitsplatz sind zu einem grossen Teil auf Infektionen und Allergien zurückzuführen, die durch die Innenraumluft verursacht werden. Da diese Krankheiten oft auf biologische Schadstoffe zurückzuführen sind, die in den Gebäuden erzeugt werden, bzw. auf eine Überfüllung der Räume, lässt sich diese Krankheitsursache signifikant begrenzen.

2. Durch eine Verbesserung der Instandhaltung von raumlufttechnischen Anlagen anfallende Mehrkosten bewirken andererseits einen Nutzen in Form eines besseren Gesundheitszustands und weniger Fehltage der Gebäudebenutzer.

3. Biologische Aerosole in Gebäuden, einschl. Privatunterkünften, entstehen vorwiegend durch anhaltende Feuchtigkeit und unzureichende Lüftung von Innenräumen und Bauelementen; solche Mängel sind durch geeignete Konstruktion und Herstellungsweise zu vermeiden.

4. Die Konzentration biologischer Schadstoffe in der Innenraumluft schwankt zeitlich und örtlich sehr stark; deshalb muss das Zahlenmaterial über die statistische Verteilung der Schadstoffkonzentrationen in Verbindung mit der Reaktion der Benutzer einen gewissen Umfang haben, um für das Risikomanagement eine Aussagekraft zu haben.

5. Die Probenahmeverfahren für biologische Schadstoffe sind im grossen und ganzen uneinheitlich. Die Verfahren für Pollen, bestimmte Bakterien und Viren werden in Bälde standardisiert; dies gilt aber nicht für Pilze, Mykotoxine und andere biologische Substanzen.

6. Labordiagnostische Verfahren für Pilze, Mykotoxine, Viren, Bakterien und anderes biologisches Material, die für die Innenraumluft evtl. von Interesse sind, sind ebenfalls noch nicht standardisiert worden.

7. Durch die Anwendung von Bioziden bei der Reinigung und Instandhaltung von Heizungs-, Belüftungs- und Kühlanlagen sowie Gebäudeflächen entstehen Risiken, sowohl direkt als auch indirekt durch das Auftreten resistenter Mikroben.

Empfehlungen

1. In Gebäuden sowie Heiz-, Belüftungs- und Kühlanlagen sollten keine biologischen Schadstoffe auftreten, die in die raumlufttechnischen Anlagen geraten können. Falls die Anwendung von Bioziden unvermeidbar ist, sollten diese nicht in Räume gelangen, in denen sich evtl. Menschen aufhalten.

2. Durch Normen und Bauvorschriften sollte eine wirksame Instandhaltung der Belüftungssysteme gewährleistet sein, indem Zugänglichkeit sowie regelmässige Inspektions- und Instandhaltungsintervalle festgelegt werden.

3. In Gebäuden, deren Benutzer nicht selbst die Belüftungsqualität bestimmen können, sollte eine verantwortliche Person bestimmt und dies den Benutzern mitgeteilt werden.

4. Um allergische Krankheiten im allgemeinen einzuschränken, sollte die Gesamtexposition gegenüber Allergenen auf ein Mindestmass beschränkt werden, indem die Allergene und ihre Entstehungsquellen in Gebäuden unter Kontrolle gebracht werden.

5. Zur Risikobewertung in Verbindung mit Allergien sollte die Expositions-/Wirkungsausprägung festgestellt werden, indem man die Antigene in der Luft und die spezifischen IgE-Antikörper in der Bevölkerung misst.

6. Mit Hilfe allgemein akzeptierter Verfahren sollten statistisch fundierte Bevölkerungsstudien durchgeführt werden, um die statistische Konzentrationsverteilung der biologischen Schadstoffe in spezifischen geographischen Bereichen zu ermitteln.

7. Die Probenahme- und Analyseverfahren für Allergene in der Luft und biologische Reizstoffe sollten standardisiert werden; ausserdem sind die Auswirkungen von Zeit, Temperatur und Feuchtigkeit zu ermitteln.

8. Das Entstehen biologischer Aerosole in Gebäuden sollte verhindert werden, indem man entsprechende Konstruktions- und Bauforderungen in die Bauvorschriften aufnimmt.

9. Das Wartungspersonal öffentlicher und sonstiger grösserer Gebäude sollte in Fragen der regelmässigen Inspektion und Instandhaltung der Gebäudeanlagen ausreichend ausgebildet werden.

10. Biologische Reizstoffe und Krankheitserreger verursachen nichtspezifische Verschlimmerungen von Atemwegs- und Hauterkrankungen und sollten deshalb durch Senkung der Konzentrationen und Einschränkung der Quellen in den Gebäuden auf ein Minimum gebracht werden.

КРАТКИЙ ОТЧЕТ

Рабочая группа в составе 19 временных консультантов из 11 стран и представителей Комиссии европейских сообществ и Европейского регионального бюро ВОЗ провела совещание в Раутаваари (Финляндия) с целью рассмотрения вопроса об объеме накопленных знаний относительно взвешенных устойчивых частиц, взвешенных аллергенов и других взвешенных материалов биологического происхождения. Внимание рабочей группы было сосредоточено на рассмотрении зданий и строительных конструкций как источников этих биологических загрязнителей воздуха и изучении результирующего воздействия этих загрязнителей на людей, живущих или работающих внутри этих помещений. Рабочая группа рассмотрела также вид, масштабы и распространение воздействия этих биологических загрязнителей на здоровье и самочувствие людей, длительное время находящихся внутри таких помещений.

Большое внимание было уделено патогенным оганизмам, размножающимся внутри зданий и строительных конструкций, таким как Legionella pneumophila, которая вызывает болезнь легионеров (легионеллез), а также болезнь легионеров нелегочного характера (лихорадка Понтиак). Такие факторы, как скученность в помещениях и использование рециркулируемого воздуха для вентиляции также могут способствовать распространению передаваемых через воздух патогенов от заболевших туберкулезом, корью, ветряной оспой и другими болезнями.

Такие компоненты вентиляционной системы, как охлаждающая башня, охладители воздуха и увлажняющие и осушающие устройства могут способствовать увеличению численности грибков, бактерий и других микроорганизмов. Эти микроорганизмы могут также размножаться на структурных компонентах здания, если относительная влажность внутри здания достигает более 70%, в то же время пылевые клещи могут размножаться в элементах отделки зданий. Такие микроорганизмы или продукты, выделяемые ими или членистоногими и более крупными живыми организмами, могут входить в состав воздуха помещений, вызывая целый ряд аллергических и раздражающих реакций у обитателей этих помещений.

Чрезмерные концентрации водных паров и сопутствующие им явления конденсации, утечки воды, выхода из строя водоотводного оборудования или непроведение очистных и ремонтно-профилактических работ - все это может способствовать попаданию в подаваемый вентиляционными системами воздух жизнестойких загрязнителей биологического происхождения. Значительная часть населения на протяжении своей жизни ощущает или может ощущать воздействие этих видов биологических загрязнителей воздуха. К числу других аллергенных и токсичных загрязнителей относятся перхоть животных, отдельные части клещей и других членистоногих, а также аэрозоли, формируемые из фекалий и мочи животных.

Комбинированное воздействие всех биологических загрязнителей воздуха внутри помещений считается одной из причин значительной доли непосещения занятий и невыхода на работу, а также значительного числа тех дней, когда активность человека ограничена. В целом среди населения 5 - 10 дней ограниченной активности на одного человека в год является нормой. Уменьшая загрязненность воздуха в помещениях биологическими загрязнителями, можно значительно снизить число острых инфекций и аллерги-

ческих реакций. Было указано на то, что стоимость
потерь в выпуске продукции в связи с невыходом на
работу и ограниченной активностью намного превышает
общую стоимость ремонтно-профилактических работ по
поддержанию в нормальном состоянии систем отопле-
ния, охлаждения и вентиляции. Из-за разделения,
как правило, обязанностей и ответственности между
организациями, занимающими помещения, связь между
этими затратами не всегда учитывается.

Выводы

1. Значительная часть заболеваний и невыходов на
работу или непосещений школы связана с инфекциями
и аллергическими реакциями, вызываемыми воздействием загрязнителей воздуха помещений. Поскольку
на эту заболеваемость часто влияет наличие биологических загрязнителей, порожденных существующими
в помещениях условиями и перенаселенностью помещений, ее можно значительно снизить.

2. Увеличение расходов, связанных с улучшением
неадекватного обслуживания вентиляционных систем,
имеет своим результатом более значительные сравни-
тельные преимущества с точки зрения улучшения здо-
ровья людей и сокращения числа невыходов на работу
и непосещения занятий.

3. Возникновение в помещениях, включая жилые помещения, биологических аэрозолей вызывается, глав-
ным образом, постоянной влажностью и недостаточной
вентиляцией помещений и строительных конструкций;
для того, чтобы не допустить возникновения таких
условий, необходимо тщательно контролировать проек-
тирование и сооружение зданий.

4. Уровни концентрации биологических загрязнителей в воздухе помещений крайне отличаются во времени и в пространстве. Поэтому базы данных о распространении определенных уровней концентрации за-

грязнителей и о соответствующей реакции людей, пользующихся этими помещениями, должны быть достаточно обширными, чтобы поставлять полезную информацию для контролирования риска.

5.   Методы отбора в природе образцов биологических загрязнителей в целом не стандартизованы. В отношении методов отбора проб на присутствие цветочной пыльцы, отдельных видов бактерий и вирусов, определенные стандарты уже почти сложились, в то время как для методов отбора проб на присутствие грибков, микотоксинов и других биологических материалов таких стандартов еще нет.

6.   Лабораторные процедуры проведения анализа некоторых грибков, микотоксинов, вирусов, бактерий и других материалов биологического происхождения, которые представляют потенциальный интерес для исследования внутренней среды помещений, пока еще не стандартизованы.

7.   Применение биоцидов в очистке и ремонтно-профилактическом обслуживании систем обогрева, вентиляции и охлаждения или для очистки и сохранения внешнего вида поверхностей в зданиях представляет собой определенный риск как непосредственного характера, так и в результате появления и распространения резистентных микробов.

Рекомендации

1.   Здания и их системы обогрева, вентиляции и охлаждения не должны становиться источниками биологических загрязнителей, которые затем попадают в вентилируемый воздух. Если применения биоцидов избежать невозможно, необходимо не допускать того, чтобы они попадали в жилые и рабочие помещения.

2.   Стандарты и строительные правила должны обеспечивать проведение эффективного ремонтно-профилак-

тического обслуживания вентиляционных систем, предусматривая возможность соответствующего доступа и регулярного инспектирования и обслуживания по установленным графикам.

3. В тех зданиях, где жители сами не могут эффективно контролировать качество подаваемого в помещения воздуха, должно выделяться лицо, ответственное за выполнение этой задачи, о чем жильцы должны быть поставлены в известность.

4. Для снижения масштабов аллергических заболеваний среди определенного контингента населения необходимо сократить до минимальных размеров воздействие на него аллергенов путем контроля за их концентрациями в помещениях и источниками.

5. Для оценки риска заболевания аллергическими болезнями необходимо определить кривые зависимости между их воздействием и реакцией, проводя замеры концентрации антигенов в воздухе и отдельных IgE-антител у населения.

6. Необходимо проводить статистически оформленные обследования населения, используя общеприемлемые методы для получения информации о распространении определенных концентраций биологических загрязнителей в специально намеченных географических точках.

7. Необходимо провести стандартизацию методов отбора и анализа аэроаллергенов и биологических раздражителей и определить влияние времени, температуры и влаги.

8. Необходимо не допускать возникновения биологических аэрозолей в зданиях, вводя соответствующие предписания в области проектирования и практики сооружения в сборнике строительных правил.

9. Обслуживающий персонал общественных зданий и деловых помещений должен проходить соответствующую подготовку для регулярного инспектирования и поддержания в надлежащем состоянии всех систем жизнеобеспечения помещений.

10. Биологические раздражители и возбудители инфекций вызывают неспецифические ухудшения болезней дыхательных органов и кожи, в связи с чем необходимо добиваться снижения до минимума их концентраций, осуществляя контроль за их уровнями в помещениях и источниками.